Josiah Parsons Cooke

Laboratory Practice

A Series of Experiments on the fundamental Principles of Chemistry

Josiah Parsons Cooke

Laboratory Practice
A Series of Experiments on the fundamental Principles of Chemistry

ISBN/EAN: 9783337176839

Printed in Europe, USA, Canada, Australia, Japan

Cover: Foto ©ninafisch / pixelio.de

More available books at **www.hansebooks.com**

LABORATORY PRACTICE

A SERIES OF EXPERIMENTS ON THE
FUNDAMENTAL PRINCIPLES OF CHEMISTRY

A COMPANION VOLUME TO
THE NEW CHEMISTRY

BY

JOSIAH PARSONS COOKE, LL. D.
ERVING PROFESSOR AND DIRECTOR OF THE CHEMICAL LABORATORY,
HARVARD UNIVERSITY

NEW YORK
D. APPLETON AND COMPANY
1891

COPYRIGHT, 1891,
BY JOSIAH P. COOKE.

TO MY KINSMAN,
FRANCIS BARTLETT, Esq.,
WHOSE SYMPATHY AND LIBERALITY
HAS ENCOURAGED AND PROMOTED THE STUDY
OF CHEMISTRY.

CONTENTS.

CHAPTER	PAGE
INTRODUCTION	5
I. DISTINGUISHING PROPERTIES	13
1. Water	14
2. Air as an Example of Aëriform Matter	43
3. Oxygen Gas	53
4. Hydrogen Gas	59
5. Sulphur	64
6. Chlorine	70
7. Carbon	73
8. Nitrogen	81
9. Magnesium and Zinc	89
10. Sodium	92
11. Copper	97
12. Iron	100
II. GENERAL PRINCIPLES	107
13. Province of Chemistry	107
14. Fundamental Laws	110
15. Compounds and Elements	113
16. Qualitative Analysis	116
17. Quantitative Analysis	121
III. MOLECULES AND ATOMS	126
18. Molecular Theory	126
19. Physical Method of determining Molecular Weights	127
20. Chemical Method of determining Molecular Weights	135

CHAPTER	PAGE
21. Conception of Atoms	139
22. Determination of Atomic Weights	144
IV. SYMBOLS AND NOMENCLATURE	149
23. Chemical Symbols	149
24. Chemical Reactions	154
25. Stochiometry	160
26. Nomenclature	162
V. MOLECULAR STRUCTURE	164
27. Quantivalence	164
VI. THERMAL RELATIONS	178
28. Heat of Chemical Action	178

INTRODUCTION.

THIS book is not intended to be used without a teacher. As far as possible, directions are given which will enable the student to perform the experiments successfully; but in many cases more precise directions and cautions are required, which should be given in the lecture room. Whenever possible the experiments should be performed by the teacher in sight of the class before they are attempted by the students, and an example of the necessary apparatus should be placed on the lecture table or in the laboratory where it can be inspected. The student ought to be left to make his own observations, and then to interpret the results with such aid as may be necessary from the instructor, who should always be present in the laboratory to oversee the work. The educational value of such a course as is here outlined depends entirely on the manner in which the work is directed and supervised. The student should be instructed, by continued reiteration, if necessary—

1. To observe the minutest particular in regard to every experiment.

2. To distinguish essential from non-essential phenomena.

3. To draw correct inferences from the results.

4. To express concisely but clearly in writing the facts observed and conclusions reached.

While the student is at work the note book should be always on the laboratory desk, and taking notes of the experiments *at the time* should be most strictly enforced. Remember that the notes are of no value except as original records of observations, and all copying of notes should be as absolutely forbidden as is the tampering with original entries in mercantile accounts. If corrections are required they should plainly appear as such.

The notes should be so written that an examiner, not necessarily the teacher, can gain from them in the shortest possible time a full conception of the work done; and this result is best secured by the use of prominent headings and paragraphs to indicate every transition in the subject matter. Remember that an examiner can not be expected to decipher hieroglyphics, and that his impression of the work will depend in no small measure on the clearness with which the results are presented.

All ciphering required in order to reduce the

results of experiments or to answer the questions in this book should be made in the note book and the purport of every calculation clearly indicated by headings. Experience shows that mere arithmetical mistakes are the most frequent sources of error in experimental work, and if, as is the too frequent practice, the separate pieces of paper on which the calculations are made are thrown away, the clue to the error is often lost. Of course, neatness in the note book is highly desirable; but this may be secured if all the odd folios are reserved for notes and the even folios for ciphering and the headings are sufficiently multiplied to prevent confusion.

It is not best to give too precise directions in regard to the note book, but to allow sufficient liberty to encourage originality both in form and in material, otherwise one note book of a class of students will be a precise transcript of every other. The following suggestions, however, will be valuable guides to a good result. In reference to every experiment—

1. State the materials taken.*

---

* In preparing the original work the author was indebted for the above rules and for other valuable suggestions to an excellent pamphlet entitled Laboratory Experiments in General Chemistry, by William Ripley Nichols and Lewis M. Norton, for the Use of Students of the Massachusetts Institute of Technology. Printed, not published, Boston, 1884. In rewriting the work for the present edition the author has further profited by the experience of the as-

2. Give a brief description, with a clear sketch of the apparatus employed.

3. State the method used, and all the circumstances of the experiment; that is, whether a substance is heated or cooled, whether a gentle or strong heat is necessary, whether an acid employed is strong or dilute, etc.

4. Describe what takes place; that is, all that is observed during and as a result of the experiment.

5. Give the conclusions drawn, or state what the experiment teaches.

The student should be given to understand clearly that experiments performed mechanically, without intelligence, or carelessly recorded, are worth absolutely nothing, and should be so estimated in any system of school or college credits.

Experiments are only of value as parts of a course of instruction logically followed out from beginning to end. In such a course there must be necessarily a great deal to be filled out by the teacher; and this can vastly better be taught from his lips, with such illustrations as he can command, than from any books. The author has therefore thought it necessary to draw up in connection with the experiments an outline of such a course

---

sistants who have had charge of the course designated at Harvard College as Chemistry B, and has received from them valuable suggestions on many points.

as he deems best suited to prepare students for the further study of natural science in Harvard College, stating the chief points to be illustrated, and giving the order in which they are best presented, leaving it to the teacher to fill out the details from his own knowledge. In rewriting the work he has also added a large number of questions and problems in order to direct attention to important inferences which might otherwise be overlooked or to enforce principles which are wont to be imperfectly apprehended by elementary students. It is intended, however, by this synopsis only to offer suggestions to the teacher, and to define more clearly the college requisition in chemistry as one of the "Advanced Studies" in the new scheme of examinations for admission.

It is thought by the writer that a course on the fundamental principles of chemistry, like the one here outlined, is far more suitable for the pupils of secondary schools than the meagre description of the scheme of the chemical elements which is presented in epitome by most of the elementary text books on this science; and in order to bring the experimental method within the means of all schools of that class, the writer has sought to adapt to the purposes of instruction common household utensils, such as may be made by a tinsmith or found at any house-furnishing store. The small petroleum or gas cooking stoves serve an admirable purpose for heating, and their ovens are excellent drying chambers; a farina kettle is a good steam bath; and the quick-sealing fruit or milk jars are not only good gas holders, but enable any student to perform experiments which formerly were made only with costly apparatus. Glass flasks, retorts, beaker glasses, glass tubing and tube apparatus, as also filtering paper, rubber tubing, rubber stoppers and corks, besides the chemicals required, are best purchased of regular dealers in chemical supplies. From these dealers may also be procured various simple devices for supporting apparatus, although equally serviceable tools can, if necessity requires, be made with blocks of wood and stout iron wire. The only apparatus of precision required, the scales and thermometers, can be imported from Germany at the prices hereafter named; and the whole cost of an outfit for a class of ten students should not exceed two hundred dollars, and much less will be required if the teacher uses a little ingenuity and can spend time in extemporizing apparatus from materials on hand. But the best apparatus will be of no use unless the teacher stands before it and speaks to his pupils out of the fulness of his own knowledge. This is an essential condition of success, and without it the experimental method should never be attempted.

In devising a course of experiments it is impossible to foresee every contingency; and the teacher into whose hands this book may come must regard the directions here given simply as suggestions to be worked out experimentally with such appliances as he can command. In many cases the directions can undoubtedly be improved, and no experiment should ever be intrusted to an inexperienced student which has not first been thoroughly tested by the teacher; and the writer is not responsible for failures which must result from a disregard of this essential precaution.

NEWPORT, *August, 1891.*

DESCRIPTIVE LIST OF CHEMICAL EXPERIMENTS

INTENDED TO ILLUSTRATE THE GENERAL PRINCIPLES OF THE SCIENCE.

CHAPTER I.

DISTINGUISHING PROPERTIES.*

MATERIAL bodies are made up of an unending variety of substances. Chemical science deals with the relations of these substances. In a world consisting of only one substance the relations of mass and energy, which are studied in physics, might be observed, but there could be no chemistry. The distinctions of substance are, there-

---

* Under the above heading, Distinguishing Properties, it is expected that the student should acquire much of the information usually presented in elementary books on chemistry in connection with the scheme of the chemical elements, but the facts should be studied in this connection simply as definite phenomena within the range of personal observation. Many teachers, mistaking the scheme of the book, not probably sufficiently elaborated in the first two editions, have practised their pupils from the first in writing the reactions of the processes employed. But this practice not only diverts attention from the main purpose of an experimental course, but so confounds fact and theory in the mind of the student that it becomes subsequently very difficult to clear up the confusion. It must not be forgotten that chemical reactions can not be intelligently written until the main features, at least, of the theory of chemistry have been fully apprehended; and that to use symbols mechanically or to learn formulæ by rote stultifies the whole study

fore, the basis of our science. In most cases these distinctions are so striking as to be obvious. No person would confound one with another salt, sugar, water, or air. But in many cases the distinctions are obscure, and can only be made manifest by careful observation and study. Hence a philosophical treatment of our subject implies a preliminary consideration of the distinguishing properties of substances, and this discussion will give the opportunity of illustrating many fundamental facts and relations which are not only of great interest in themselves, but which will also serve as the basis of a knowledge of the principles of chemical science.

1. Water.

EXPERIMENT 1. *Density of Water.*—A closed cylinder, about 5 centimetres long by $2\frac{1}{2}$ centimetres in diameter, of sheet tin or brass, with a hook at the top, which can be made by a tinsmith, is required for this experiment. Before soldering on the top the cylinder should be loaded with lead shot, so that it will sink in water. A pair of scales, with a set of weights,

---

of physical science. To make evident the effect of ordinary textbook teaching the following question was asked for a series of years in the examination papers on chemistry for admission to Harvard College: "On what evidence is our knowledge of the composition of water based?" and in at least four cases out of five the answer given was: "Water consists of hydrogen and oxygen, because $H_2 + O = H_2O$."

is also required. Hand scales, with horn pans and brass beam 20 centimetres long, and with a set of weights from 0·01 to 100 grammes, are well adapted for all the experiments here described, and cost in Germany only $2.37. The student first measures the size of the cylinder. This may be done by fitting exactly a sheet of glazed paper to the convex surface, and measuring with a millimetre scale the length and width of the paper. The volume of the cylinder, in cubic centimetres, can now be calculated, and the student should be shown how to estimate approximately the probable error of his result. The cylinder is next to be weighed, first in air, and secondly under iced water. For this purpose the scales are best hung by a cord passing over a pulley and secured to a belaying-pin, so that they can be adjusted at any height required. The cylinder is best suspended beneath the pan by a silk thread, which plays freely through a hole made for the purpose at the centre of the pan, and is hung from the same hook as the pan itself. The difference between the weight of the cylinder in air and its weight under water will now give the weight of a volume of water equal to that of the cylinder, and from this value the weight of one cubic centimetre of ice-cold water is easily found. Since, by definition, a gramme is the weight of one cubic centimetre of pure water at 4°

C. (the same as that of ice-cold water within the limits of accuracy of student's work), the result should be closely one gramme. The probable error of the work may now be estimated by comparing the several results of as large a number of different students as possible. Such a comparison may be made very instructive by writing the results, distinguished by numbers or otherwise, in a vertical column on the blackboard, and, after finding the average value, placing opposite to each result the difference between it and this average value. It will now probably appear that some of the results are far astray, in consequence of careless work or mistakes in calculation. These should be thrown out, a new average value taken, and a still closer scrutiny applied, when, on arranging the remaining results in the order of their values, each will be found to differ from the next by a small and nearly constant quantity. The final average must represent very closely the best result that can be obtained with the imperfect means employed; and if it differs by more than a few milligrammes from one gramme, there must be some constant source of error which the teacher should seek to discover. Such a comparison as this not only furnishes an unimpeachable test of the relative skill of the different men in a class of experimenters, but also gives a clearer idea of the necessary limitations of experimental methods than can be acquired in any other way, and it is all-important that students should gain a clear idea on this point from the start.

NOTES, QUESTIONS, AND PROBLEMS.*

(1) What is the principle of Archimedes, and by what simple consideration can you show that it must be true?

(2) How can you apply this principle to determine the specific gravity of a solid body?

(3) In this book by specific gravity is always meant the ratio between the weight of a substance and that of an equal volume of water or that of some other standard material, and by density is always to be understood the weight of the unit volume of a substance. In the French system the density of a substance is the weight of a cubic centimetre of the material in grammes. In the English system it is the weight of a cubic inch of the material in grains. Specific gravity, then, is a ratio, but density is an absolute weight. If we multiply the density of the standard of reference (that is, the weight of the unit of volume) by the specific gravity of a substance we shall have the density of that substance—that is, $\delta = \delta'$ Sp. Gr. Further, if we have given the volume of a body and the specific gravity of the material of which it consists, the weight of such body must be equal to the product of the weight of one unit of volume (its density) into the number of units of volume, $W = \delta\, V = \delta'$ Sp. Gr. V, in which δ' is the density of standard of reference as above. In the French system, since by definition the gramme is the weight of one cubic centimetre of water, it is true when the standard of reference is water that

$$W = \text{Sp. Gr. } V,$$

W standing for a certain number of grammes and V for a certain number of cubic centimetres. With any other sys-

---

* This heading will not be repeated, but must be regarded as applying to the numbered paragraphs in small type, which will very constantly follow the descriptions of experiments.

tem of weights and measures we must retain in the formula the value δ' when

$$W = \delta' \text{ Sp. Gr. V.}$$

In the English system W stands for weight in grains, V for volume in cubic inches, and δ' for the weight of one cubic inch of water in grains. In all systems Sp. Gr. $= \frac{\delta}{\delta'}$. In the French system alone where $\delta' = 1$ gramme we have Sp. Gr. $= \delta$. This equality obviously arises solely from the selection of a cubic centimetre of water as the unit of weight' but while this assumption greatly simplifies many calculations, and is one of the chief merits of the French system of weights and measures, it has led to a confusion of ideas as well as of terms which it is important to keep distinct.

(4) In the original scheme for the French system of measures and weights, the metre was defined as the one ten-millionth part of the quadrant of a meridian of the earth, and, in order to construct a bar corresponding to that definition, French engineers actually measured an arc of the meridian which passes through Dunkirk, in northern France, and Barcelona, in Spain, determining with extreme care by astronomical means the latitude of these two places, so as to find what ratio the arc measured bore to the quadrant. As in all geodetic measurements on a large scale, the end points were connected by a system of triangulation, referred to certain base lines actually measured at convenient localities by means of bars of standard length. These bars were necessarily graduated to the measures of length previously in use in France, and as the result of the work the length of the quadrant was found (approximately) in terms of the old standard called a toise, and then there was only the mechanical difficulty left of making a metallic bar equal to a certain fraction of a toise, and this was the new standard since called the metre. A metre rule having been thus constructed with the subdivisions accurately marked upon it, the weight of one cubic centimetre of water (the new unit of weight) was then found by a method in all respects similar to that employed in the above experiment. A metallic cylinder was made much larger than those here used, and care-

fully wrought in a lathe so as to be as nearly as possible of uniform dimensions. These dimensions were next measured with the metric scale with extreme accuracy and the volume of the cylinder in cubic centimetres calculated. It was then only necessary to weigh the cylinder in the air and under water, when the chief data were obtained for calculating the weight of one cubic centimetre of water the required gramme. We say the chief data because all the weights and measurements had to be corrected for variations in temperature or in other conditions which it would be out of place to discuss here. But the result of all was to give the weight of the new standard in terms of the old system of French weights used in the work. In few words, it was found that a gramme equalled so many old French grains, and afterwards standard gramme weights were made in brass or platinum to correspond to the value thus found.

In order that the student may fully understand these somewhat complex relations, he should repeat the above experiment with English weights, and thus find the weight of a cubic centimetre of water in English grains, when he can himself adjust a gramme weight with a piece of sheet lead. If this is not possible from the want of an accurate set of English weights, he may be set a problem in this form: Given a cylinder of such dimensions (in centimetres), weighing so many grains in air and so many under water, what is the value of a gramme in grains?

1 gramme = 15·432 grains.

Ex. 2. *Expansion of Water by Heat.*—Provide a cylindrical glass bulb about 20 millimetres wide and 30 millimetres long opening into a tube about 2 millimetres wide and 200 millimetres long. In order to facilitate the emptying and drying of the bulb, a short tubulature of the same size as the stem should be provided at the opposite end and drawn out so that it may be readily closed by melting the glass or opened by breaking off the tip as required. Such a bulb tube will have the form of a measuring instrument well known in chemistry as a pipette. The tubulature being closed, the bulb tube is first filled with colored water (freed from air by boiling in an open beaker glass) to a point about 30 millimetres above the neck. To do this, first heat the empty bulb in a free flame (but not hotter than the hand can bear), and then dip the open end of the stem into the colored water. Wait until a teaspoonful has been drawn into the bulb, and then bringing the stem upright boil carefully the water in the bulb until the interior is full of steam; then, grasping the bulb with some holder to protect the hand, again plunge the mouth of the stem in the colored water, kept boiling meanwhile. The experiment requires a little dexterity, and is for that very reason good practice. If the manipulation is successful the bulb tube will completely fill with water without showing the smallest air bubble. It can be set away upright when still nearly boiling hot, and when cold the water will be found to have descended in the stem to about the required amount, provided always the dimensions given have been followed in the construction of the bulbs.

The filled bulb is now to be packed in broken ice and the level to which the liquid sinks in the stem marked (most readily with the burnt end of a match). It is next to be heated in a steam bath and the level again marked. (The bulb must not rest on the overheated bottom of the steam bath, else large fluctuations of level will be noticed, caused by the formation of steam in the interior.)

In order now to measure the amount of expansion between the freezing and the boiling point the bulb tube should first be emptied, breaking off the tip of the tubulature for the purpose. Using now the instrument as a pipette, water should be sucked into the bulb and tube until the curved surface (meniscus, so called) is tangent to the first mark. This water should then be run out into a small tared beaker and weighed. Again the bulb and tube should be filled but to the upper mark, and then the column of water between the two marks run out with the greatest possible care into a small tared stoppered bottle and weighed if possible to a milligramme. It is best to use pure distilled water for the purpose at the temperature of the laboratory, and it will require a little experience to run out the exact amount of water between the marks. But repeated trials can be made until the result is obtained, and it will be found that by moistening the finger which controls the mouth of the stem great sharpness can be secured.

As the two weights thus obtained are to each other, so is the volume of water at the freezing point to the increase of volume at the boiling point. Hence it will be easy to find the fraction of its volume which one cubic centimetre of water would expand between these temperatures, and this is technically called the coefficient of expansion. This value can be only approximately found in this way; but, as in the first experiment, so here and in all subsequent experiments involving quantitative measurement the probable error with the tools used should be estimated, although the point may not again be referred to.

This experiment illustrates not only the expansion of water by heat and its contraction when cooled, but also the expansion of air by heat, as shown in the method of filling the bulb. The fact that it is the relative and not the absolute expansion of water which is measured should be fully explained. The bulb tube should be kept for another experiment.

(1) The coefficient of cubic expansion of glass (that is, the expansion of one cubic centimetre in volume between 0° and 100° C.) is 0·0025. What is the absolute expansion of water according to your experiment?

(2) Why is it important to determine the smaller of the two weights required so much more sharply than the larger of the two?

(3) Why boil so thoroughly the water with which the bulb tube is first filled?

Ex. 3. (a) *The Melting and Boiling Points of Water.*—For this experiment thermometers with a scale on the tube from —5° to 360° (such as are sold in Germany for eighty cents) are required. The student should first test the melting point of ice, repeating the observation several times with different amounts of ice and under different conditions, until he gains a clear idea of the constancy of the thermal state at which the change takes place. He should also repeat the observation with different thermometers, using if possible thermometers of short range, with the degrees divided into tenths; and the cause of what is called the rise of the zero point should be explained. Next, the boiling point of water should be observed, and attention should be called to the constancy of the thermal condition at which this phenomenon takes place under the same atmospheric pressure; and the small variations which depend on changes of the barometer should also be explained, so far as the previous knowledge of the student will permit. The chief principle to be taught by this experiment is the constancy of the thermal conditions which we call the melting point of ice and the boiling point of water.

(1) Does the quantity of ice melting or the quantity of water boiling make any difference in the temperature of these materials as indicated by the thermometer? What inference do you draw from these facts?

24   LABORATORY PRACTICE.

(2) How does the temperature of the ice compare with that of the lambent water, or the temperature of boiling water with that of the steam above it? What becomes of the heat that enters the containing vessel?

(3) On what evidence do you infer that the freezing and boiling points of water are constant under the same conditions?

(4) What is the effect of a change of atmospheric pressure on the boiling point?

(5) It is often the case when a thermometer is packed in broken ice that the mercury column does not stand exactly at the zero point of the scale. What is the cause of this anomaly, and ought observations made with the instrument to be corrected therefor? When the thermometer is dipped in water to which an inadequate amount of ice is added the thermometer seldom marks the true zero. Why is this to be expected?

(*b*) *Distillation of Water.*—Use for the purpose a glass retort holding about two hundred cubic centimetres. First try to distil without any provision for cooling the neck of the retort. Afterwards wrap the neck loosely with several folds of filtering paper, so as to form a covering (several inches shorter than the neck) secured at the two ends with rubber bands. Adjust a hose connected with a tap so that water will trickle through a hole made for the purpose near the upper edge of the cover, and lead off the stream from the lower end with a line of lamp wicking that has been previously soaked in water. Repeat now the experiment.

(1) What is the material above the water while boiling in the retort? Is this material the same as water? Has it

weight? Is there necessarily any loss of weight in the process of distillation? When water boils away in an open kettle what becomes of it?

(2) Why is it necessary to adopt some means of cooling the retort neck in order to distil rapidly?

Ex. 4. *Construction and Principles of the Thermometer.*—The student may next make, with the tube apparatus of Ex. 2, a water thermometer, and compare it with one of the ordinary mercury thermometers used in the laboratory. For this purpose the tip of the tubulature must be re-sealed and the bulb filled with colored water with the same precautions as before. The fixed points having been permanently marked (best with a file wet with kerosene), the interval between them should now be divided into twenty equal parts, corresponding on the mercury thermometer to 5°, 10°, 15°, 20°, etc., and a pasteboard scale adjusted to the stem. The student should next compare the two thermometers, dipping them for the purpose side by side and gradually raising the temperature of the bath.* Let him note the heights of the water column corresponding to each one of the series of degrees of the mercury thermometer just mentioned, and mark each of these positions

---

* On account of the much larger mass of the materials to be heated and the much greater capacity of water for heat, the water thermometer always lags behind the mercury thermometer, and in order that the comparison should be just the temperature must be held at each point long enough for an equilibrium to be established.

on the scale. After the positions have been marked with a pencil, the scale should be neatly drawn, with the equal divisions on one side and the irregular divisions on the other side of the stem.

(1) The significance of this experiment will not be appreciated unless the student has himself made a mercury thermometer, or unless its construction has been fully explained by the teacher. There are difficulties in the construction which do not recommend it as a general laboratory experiment, but it may safely be entrusted as a supplementary experiment to the more skilful men. Bulb tubes should be provided for this purpose with a cup at the mouth of the stem to hold the mercury required to fill the bulb, and inexperienced manipulators will be most successful with very small bulbs (not holding over one fifth of a cubic centimetre) with proportionally fine tubes.

(2) If equal intervals on the scale of the mercury thermometer represent equal changes of temperature, can the same be true of the water thermometer?

(3) Using only the second scale of the water thermometer with irregular intervals constructed as described above (that is, adjusted to each of the divisions 5°, 10°, 15°, 20°, etc., of a mercury thermometer), it is obvious that the water thermometer would give the same indications as the mercury thermometer; but how would the lengths of the successive divisions of this scale vary among themselves between 0° and 100°?

(4) What must be the cause of the great discrepancy between the two thermometers, when both are graduated with regular intervals? and why is the mercury thermometer to be preferred as a measure of temperature?

(5) Do the intervals of the scale of the mercury thermometer correspond to equal differences of temperature? And what must be the nature of the thermometric material of which this would be true? Could it be true of any material in a glass envelope? How is it with air?

(6) State your conception of the term temperature as formed from this experiment, and from the considerations suggested by the above questions.

(7) It is not expected that the student can fully answer all these questions from his previous knowledge. They are intended to stimulate thought and suggest to the teacher directions in which the subject may be developed. The student can at least be made to comprehend that temperature is a *thermal condition* of which equal intervals can be predicated and measured with close approximation, although the condition itself can only be defined theoretically.

Ex. 5. (a) *Irregular Expansion of Water near its Freezing Point.* — Having packed in broken ice the water thermometer described under the last experiment, wait until the column is stationary, and then raise the instrument from the ice and watch the motion of the column as the temperature rises. Note that the column sinks to a perceptible extent before it begins to rise, but afterwards expands steadily with the increasing temperature. The experiment is intended to show that the point of the maximum density of water is above the freezing point, and in connection with it the relations of this remarkable property of water in the economy of nature should be explained to the student.

(b) *Conduction of Water for Heat.*—Select a narrow but long test tube, nearly fill with water. Grasping the tube at the bottom, apply a flame near the top of the tube a few centimetres from the surface of the water until the liquid boils.

Repeat the experiment, applying the flame near the bottom of the tube while holding it at the top.

(*c*) *Conduction Currents.* — Repeat the last phase of the experiment, adding some shreds of filtering paper to the water, whose motion will indicate the play of the currents.

(1) Why can water be most readily heated by applying the flame at the bottom of the containing vessel?
(2) Why in winter do the ponds only freeze on the surface?
(3) Would a water thermometer made as above show the exact point of maximum density?

Ex. 6. (*a*) *Density of Ice. First Method.*— Float a lump of ice of regular shape on water and observe as nearly as you can estimate with the eye what fraction of the volume is immersed. Mix now about an equal volume of alcohol with the water until the ice neither floats nor sinks, and then, removing the ice, take the specific gravity of the mixture by means of the tin cylinder of Ex. 1.

(*b*) *Second Method.*—Provide a tin cylindrical vessel of the capacity of about 250 cubic centimetres, which should be packed round with ice and salt like an ice-cream freezer (use glass beaker 500 cubic centimetres for outer vessel); provide a second, similar in every respect, but two thirds filled with kerosene. Suspend now a piece of ice weighing about twenty-five grammes to the beam of the balance by means of a silk thread, as in Ex. 1, and adjust so that the ice shall hang within the first cylinder, and weigh the ice. Replacing then the first cylinder with the second, weigh the ice immersed in kerosene. Lastly, weigh, immersed in kerosene, the metallic cylinder described in Ex. 1. We have now four weights—the weight of ice in air, the weight of the same ice immersed in kerosene, the weight of the metallic cylinder immersed in the same kerosene, and from Ex. 1 we have the weight of this cylinder immersed in ice-cold water. From these data we can easily calculate the specific gravity and density of the ice.

(1) The second method requires more care as well as more apparatus than the first, and may be reserved for the more skilful experimenters.

(2) What fraction of an iceberg is immersed in the ocean? Specific gravity of sea water, 1·03.

(3) A block of ice weighs 36·72 kilogrammes. What is its volume?

Ex. 7. *Expansion of Water in Freezing.*—Provide glass bulb tubes similar in construction to those described under Ex. 2, only having larger stems, about three millimetres in diameter. Fill the bulb about one half with water and boil the liquid to expel the air. Fill the rest of the bulb and a small portion of the stem with kerosene. (This is easily done with a small tunnel, the neck of which has been drawn out into a long fine tube.) Next freeze the water by immersing the bulb in a mixture of ice and salt. To prevent breaking the glass, hold the stem obliquely and keep the bulb turning while the water is freezing. Raise the tube from the freezing mixture and follow the descent of the column as the ice melts. The motion of the column is somewhat erratic, owing to the circumstances that the water, kerosene, and glass expand independently and at very different rates, and that the heat diffuses through those materials only slowly; but the main feature of the experiment far surpasses all the lesser effects. Observe all variations as closely as possible and seek to explain them.

(1) The volume of water used in the bulb can readily be found by weighing the glass before adding the water and after the water has been boiled and cooled. The expansion of the water in freezing can then be measured by following the same general method described under Ex. 2 and the result compared with those of the previous experiment. This phase of the experiment may serve when occasion offers as an extra exercise.

(2) Why boil the water before freezing?

(3) Any attempt to measure the density of steam would be premature in this connection, but the general result of such measurements should here be stated and illustrations given of the great expansion which the conversion of water into steam involves The specific gravity of steam is 0·6235 referred to air at the same temperature and pressure, and one cubic centimetre of boiling water yields approximately 1,627 cubic centimetres of free steam at the normal pressure of the air (one cubic inch yields nearly one cubic foot). In a

closed vessel partially filled with water—like a steam boiler—both the density and pressure of the steam which forms in the assumed empty space above the liquid vary with the temperature, increasing rapidly as the temperature rises, according to a complex law. Tables giving the density and pressure corresponding to successive temperatures are to be found in works on physics, and as these values have important applications even in this elementary book they should be shown and explained. When high-pressure steam escapes from a boiler into the atmosphere, the steam expands until its pressure is reduced to that of the atmosphere, and then mixes with the air. A locomotive engine is constantly pumping steam into the atmosphere, and at every revolution of the driving wheels each cylinder is filled and emptied twice. In a run of twenty miles a very large volume of steam is thus spent, and is supplied by the evaporation in the boiler of a comparatively small amount of water.

Ex. 8. *Capacity of Water for Heat.*—Provide a round pasteboard box; line the inside with felt at least two inches thick, covering both the bottom of the box and the under side of the cover, but not the rim, which should be at least three inches wide; cover the felt with a second lining of thick paper. Procure also a cylindrical vessel made of the thinnest sheet brass which can be obtained. This vessel should be ten centimetres in diameter and fifteen centimetres high, with thin flanges on the sides and bottom. It is to stand inside the felt-lined box, from which it is kept apart by the flanges, and the dimensions should be such as to leave half an inch of air space around the brass vessel. It should be provided with a stirrer, made also of thin sheet brass, like a turbine wheel; and there must be a hole through the felt-lined cover of the outer box large enough to pass the tubular axis of the stirring wheel, and the thin brass tube which forms the axis must, in its turn, be large enough to pass the bulb of a thermometer. This apparatus is called a calorimeter, and will be used for many experiments.

Weigh out in the brass vessel about 500 grammes of water, noting the exact weight. Replace in the calorimeter, and wait until the temperature is constant, as indicated by a thermometer graduated to one tenth of a centigrade degree. Weigh out next in a dipper, or some other vessel with a handle which can conveniently be heated in a steam bath (a farina kettle, for example), about 500 grammes of small iron nails. When the metal has fully reached the temperature of the steam (and this will be most readily secured by covering the top of the kettle with a towel) pour the nails as quickly as possible into the water, which may be left uncovered during this experiment. Follow now the temperature of the water, and note the highest point reached, taking care to secure uniformity by moving the stirrer several times up and down before each reading of the thermometer. The temperature will rise only a few degrees; for the quantity of heat given out by the metal in cooling from

100° to the temperature of the calorimeter is only sufficient to raise the temperature of the water by a comparatively small amount, showing that the capacity of the water for heat is far greater than that of the iron nails. Make now a similar experiment with granulated copper or with brass turnings, a third with lead shot, and a fourth with glass beads. By means of the data thus obtained the student should calculate the capacity of iron, copper, lead, and glass for heat, as compared with water. Adopting for the unit of measure that *quantity of heat which will raise the temperature of one gramme of water one centigrade degree*, the student will find what fraction of a unit of heat will raise the temperature of a gramme of copper, of iron, of lead, or of glass one degree. These values are called the specific heats of the substances. It is expected that the student will gain through these measurements a clear conception of the difference between temperature and quantity of heat; and this is the most important point here illustrated. After the student fully understands how temperature is measured and how quantity of heat is measured, he will be easily able to construct the simple formulæ by which the ordinary problems connected with specific heat are solved. These problems should be multiplied until the subject is fully mastered. The student will thus come to appreciate how very great is the capacity of water for heat, and he should be shown the important relations of this storage capacity in the economy of nature.

(1) It is an obvious caution in this experiment to prevent the calorimeter from being affected by radiation from the steam bath.

(2) Make clear the distinction between the unit of heat and the unit of temperature.

(3) How many units of heat would be required to raise the temperature of 100 grammes of brass from 18° C. to 23° C.? How many grammes of water would be the thermal equivalent in this respect of 100 grammes of brass?

(4) Must not the brass vessel of the calorimeter have an influence on the results of the above experiments? How can you find its thermal equivalent and correct your results therefor? Note that this correction must always be made in all similar cases, as in the two following experiments.

(5) Why are insular climates comparatively moderate?

Ex. 9. *Latent Heat of Water.*—Prepare the calorimeter as in Ex. 8, and note the weight and temperature of the water. Stir in finely broken ice so long as it promptly melts. Close now the the calorimeter and note the fall of temperature. Lastly, remove and reweigh the brass vessel; and this weight, when compared with the first weight, will give the amount of ice added. If no heat were required to melt the ice, it is obvious that its effect on the calorimeter would have been the same as that of an equal quantity of ice-cold water. Bearing this in mind, it will be easy to calculate from the experimental data how many units of heat are required to melt one gramme of ice. This is called the latent heat of water. Explain the nature of this phenomenon and the objections to the use of the term "latent" as applied to it.

(1) Take care that the broken ice is as dry as possible and not drenched with running water. Why?

(2) How many grammes of broken ice would be required, when stirred into 500 grammes of water at 20° C., to reduce the temperature of the liquid to 0° C.? Why should you expect that practically more than the theoretical amount would be required?

(3) In what sense can freezing be regarded as a warming process?

Ex. 10. *Latent Heat of Steam.*—Prepare the calorimeter as before, and pass into the water for a few minutes a current of dry steam. The steam may be drawn from the steam pipes of the laboratory through tubes of thin sheet brass, which, if first warmed by passing the current for a few moments before dipping the mouth of the tube under the water, will deliver nearly dry steam. When the laboratory is not heated by steam the steam may be generated from a glass flask; but this must be so screened as not to affect the calorimeter. Stir, and observe the rise in temperature. By reweighing the brass vessel the amount of steam condensed is determined; and by applying the same course of reasoning as in the last experiment the student can easily calculate the amount of heat developed when one gramme of steam condenses to one gramme of boiling-hot water without change of temperature. We thus measure the latent heat of free steam—that is, of such steam as rises from water boiling under the ordinary pressure of the atmosphere, which necessarily has the same tension as the atmosphere and the same temperature as the boiling water. The teacher may here add that the latent heat of water changes with the temperature according to a well-known law. Let him also consider in this connection the use of steam in heating, and also the effect of the aqueous circulation in modifying the temperature of the globe.

(1) Why must the steam be dry?
(2) Why the necessity of cooling the neck of the retort in the experiment on the distillation of water?
(3) Why do the rain storms tend to equalize the climates of the earth?
(4) In many industrial processes water is boiled in wooden tanks by blowing steam into the liquid. In such a case, if we start with 1,000 kilogrammes of water at 15°, how much liquid (condensed steam) will be added in raising the temperature to the boiling point? It is here assumed that there is no loss, but as the water nears boiling a great deal of steam escapes. How does this loss affect the result?

Ex. 11. *Solvent Power of Water.*—1. Weigh into a test tube ten grammes of water. Weigh out successive portions of one gramme each of cupric sulphate. Add the first portion to the test tube, cork tightly, and shake; and if this dissolves, add the second, and so on until, after continued shaking, the last portion fails to dissolve. Estimate the number of parts of the salt that have dissolved in one hundred parts of water. In like manner test the solubility of potassic nitrate, potassic sulphate, acid potassic tartrate, and baric sulphate.* If even the first portion fails to dissolve, ascertain whether any has dissolved. This is best done by filtering some of the mixture and evaporating a few drops of the filtrate (which should be perfectly clear) on a strip of window glass.

2. In the same way experiment with common salt (sodic chloride), baric nitrate, and crystallized sodic sulphate; but when each solution has become saturated at the temperature of the room, warm slowly, and as often as the salt is entirely dissolved add a new portion of one gramme, finally bringing the liquid to boiling. Carefully observe the effects. What does the experiment show? What effect may the heat of the hand produce in the examples under 1?

Points to be made prominent: The very general, but limited, solvent power of water; effect

---

* In cases like this, where a repetition of the same experiment would be tedious, and the additional practice not important, the end will be gained by having the determinations under different conditions made by different students, and the results compared before the class.

of temperature on the solvent power; features exhibited by different substances, both as regards the extent of their solubility and its variation with the temperature; the conception of a saturated solution.

(1) When hot water dissolves more of a salt than cold, what should you expect would follow on cooling a hot saturated solution? Try the experiment with nitre and copper sulphate.

(2) If water holds a non-volatile material in solution, what should you anticipate would follow on evaporating the water? What if the material were in itself volatile? Try the experiment by evaporating on a glass plate a few drops of solution : (1) of common salts, (2) of chloride of ammonia, and (3) of aqua ammonia, using in each case weak solutions and heating cautiously on iron plate over a flame.

Ex. 12. *Hard Water.*—Boil down in a clean saucepan half a litre of well water to a few cubic centimetres. Transfer to a tared porcelain crucible, and, after evaporating to dryness, weigh the residue and try to recognize the chief ingredient by the taste. Calculate the per cent of solid impurities dissolved in the well water. Redissolve the residue as far as possible in a few cubic centimetres of water, and shake up in a test tube with a solution of soap, adding the soap solution in small successive portions. Distil another portion of the same well water, and test the distillate by evaporating a few drops on a glass plate, and with soap solution as before. Compare the results.

Ex. 13. *The Crystallizing Power of Water.*—

Prepare a saturated solution (about 20 grammes each) of alum, potassic ferrocyanide, potassic nitrate, sodic nitrate, ferrous sulphate, and cupric sulphate. Pour each solution into a shallow dish, and, protecting it from dust with a cover of porous paper, leave the solution to evaporate in a warm, dry place. Examine from time to time, and study the forms obtained. This work may be divided.to advantage among several students, who may be shown how to "nurse" the crystals, and will compete with each other to obtain large and perfect forms. In this connection the fundamental forms of crystals should be studied, and the production of natural crystals in geodes and mineral veins explained. We have selected one example from each system of crystals; but the teacher may multiply these examples according to the material at his command, so as to illustrate all the characteristic features of crystalline growth.

(1) May not crystals be obtained as readily by cooling a hot saturated solution as explained above. What advantage is to be gained with the slower process recommended here ?

(2) Describe the six types of crystals exhibited by the substances above selected.

(3) Are these crystalline forms accidental, depending on external relations, or are they qualities, and therefore characteristic of the substances used ?

Ex. 14. (*a*) *Water the Medium of Chemical Changes.*—Mix in a mortar half a gramme of pulverized dry sodic bicarbonate with the same weight of pulverized dry tartaric acid. Transfer to a test tube and pour on water. After the action has subsided evaporate the liquid to dryness and compare by tasting the residue with the materials used.

(*b*) Mix in a mortar a few milligrammes of dry pulverized baric chloride with about the same amount of dry pulverized sodic sulphate and observe the effect. Weigh out now as accurately as possible 208 milligrammes of the first salt and 142 milligrammes of the second. Dissolve each in separate test tubes in about five cubic centimetres of water. Heat the solutions nearly to boiling, and pour the first into the second. When cool enough to handle, shake the materials together and leave to settle. Pour off (decant) a portion of the clear liquid and evaporate it to dryness. Taste the residue and compare with the substances taken.

(1) Why weigh so accurately the materials used?
(2) The chief point to be noticed in the above experiments is that the dry powders are perfectly inert towards each other and that no change takes place until water is added. In saying that water is the medium of the change we mean that it acts chiefly in virtue of its solvent power, although it often happens (as will afterwards appear) that a portion of the water present may enter into union with the product formed or may be separated from the materials as one of the results of the process. This concurrence of water in chemical changes is so universal as to be one of the most

important features of such processes; and as it very frequently obscures more essential phases, the student should become acquainted with the general facts from the start.

(3) The solid precipitate which falls in (*b*) (baric sulphate) is obviously insoluble in water, and when materials are brought together in solution there is always a tendency to such a transfer of their constituent parts as will produce insoluble compounds. The effect in this case is one of wide significance and important application.

Ex. 15. *Water in Combination.*—Heat some small bits of the mineral gypsum at the bottom of a closed glass tube held obliquely, and seek to recognize the clear liquid drops which condense on the walls. Heat now a weighed amount of gypsum in a porcelain crucible, and from the weight of the residue determine what per cent of water gypsum contains. Make similar experiments with crystallized baric chloride. Try also, for comparison, an experiment with common salt, using only one gramme of finely pulverized material and covering the crucible to avoid loss by snapping. As illustrating the same points, take next a lump of quicklime weighing about fifty grammes, and, having noted the exact weight, place it in a capacious evaporating dish previously tared. Now pour upon it water little by little so long as the liquid is absorbed and weigh the product. From these weights it will be seen that water has united with the lime. For the last illustration, mix with water to a thin paste some plaster of Paris (dried gypsum), and pour the plaster over a silver dollar previously oiled and placed on a sheet of oiled paper. When the plaster has hardened remove the cast. In this connection the teacher should define the term "water of crystallization" and make the distinction between analysis and synthesis.

(1) How can you prove that the liquid drops above mentioned are water?

(2) Which of the above processes are analysis and which synthesis?

Ex. 16. *Water itself a Compound.*—At this point the student should be shown the decomposition of water by an electric current and its subsequent synthesis with the endiometer in order that he may clearly see that our knowledge that water is composed of oxygen and hydrogen gases rests on evidence of the same kind as that which appeared in the previous experiment.

These experiments can not be performed by the student himself without more expensive appliances than most schools can command, and are best shown to the whole class at once on the lecture table. The writer would here state that in his opinion it is not necessary, in order to secure the full advantages of the experimental method, that each student should perform every experiment for himself. Indeed, if this is attempted, a course in chemistry must either be made very meagre or very expensive. If the student actually performs in the laboratory a sufficient number of experiments to give him the spirit of the method, he will usually comprehend the full significance of those which are plainly exhibited on the lecture table by the instructor.

2. Air as an Example of Aëriform Matter.

Ex. 17. *Air has Weight.*—Select two flasks of about 250 cubic centimetres capacity which have been blown in the same mould and are therefore of equal size. Fit them tightly with corks. Cork one permanently and reserve it as a counterpoise. Add to the other about twenty-five cubic centimetres of water and boil the water over a lamp until the interior is full of steam; then cork, remove the lamp, and allow to cool. Tare now the second flask with the first and such additional weight as may be necessary. Draw the cork from the second flask, and after air has entered determine the increase of weight. Determine the volume of the air weighed by adding water from a graduated measure until the flask is filled to the former level of the cork. From these values the weight of one litre of air under the conditions of the experiment can be roughly determined.

(1) Accurate methods of determining the weight of aëriform substances are beyond the reach of elementary students, but the values obtained for a few of the gases referred to in this book may here be collected for reference.

Weight of one litre when $H = 760$ millimetres and $T = 0°$. Also specific gravity when air $= 1$:

	Weight.	Specific Gravity.
Air	1·2932	1·
Oxygen	1·4303	1·1056
Nitrogen	1·2562	0·9713
Hydrogen	0·0896	0·0692
Carbonic dioxide	1·9775	1·5291
Nitrous oxide	1·9746	1·5269

(2) Why is it so much more difficult to determine the weight of a mass of gas than that of a solid or liquid body ? Is it a definite quantity ?

(3) What is the use of the corked flask as a part of the counterpoise ?

Ex. 18. *Relation of Volume to Pressure. (Law of Mariotte.)*—Required a graduated glass tube about 300 millimetres long, divided into fifths of a cubic centimetre, open at the lower end and closed at the upper by a tubulature guarded by a pinch cock like a Mohrs burette, also a piece of plain glass tubing of about the same dimensions as the first. Connect the two with a length (about 300 millimetres) of stout rubber tubing and wire the ends on to the open mouths of the glass tubes and then support with clamps the two tubes side by side, the rubber connecting tube hanging in a curve below (the last must be stout enough to prevent kinking). Open now the tubulature and pour mercury into the open mouth of the side tube until the level of the two columns (which will be the same if there is no obstruction) stands about midway of the graduation. Close the tubulature and read the volume of the confined air; also read the barometer, which gives in millimetres of mercury column the pressure to which the confined air is exposed. Next raise or lower the side tube to as great an extent as the apparatus will allow, but only a small amount at a time, and in each position measure with a graduated rule the difference in the heights of the two mercury columns in millimetres. Read the corresponding volume of the air, and also take again the height of barometer if there has been any change. Add to each difference of level (positive or negative) the corresponding height of the barometer. Call this sum H, which stands for a number of millimetres. Make out a table giving on one side the observed volumes, V, in cubic centimetres, and on the other side in a parallel column the values of H in millimetres. Look now for a relation between the quantities thus tabulated, and it should be found that in each case

$$V : V' = H' : H.$$

(1) A Mohrs burette can be used for this experiment, only as it is graduated the wrong way the value of the space between the lowest division and the stop cock must first be measured and added to the readings reversed.

(2) In handling the apparatus the student must be very careful not to heat the measuring tube, but give time before reading for the whole to come to the temperature of the room, which is assumed to be constant during the experiment.

(3) This experiment is not only calculated to give a clear

conception of Mariotte's law, but it also furnishes important practice in the measurement of gas volumes and gas tension. It assumes a knowledge of the use of the barometer and of the fundamental principles of pneumatics, and, like these subjects, might be relegated to the course on physics. But the whole subject is of such fundamental importance in the theory of chemistry that the chemical student can not fail to profit by the exercise, even if it is a review of previous work.

(4) What is meant by the tension of a gas, and how is it measured? Is there any distinction between tension and pressure? What is the standard pressure to which all measurements of gas volumes should be reduced?

(5) A volume of air was found to be 200 cubic centimetres. The barometer at the time stood at 740 millimetres. What would have been the volume if observed when the barometer stood at 760 millimetres. *Ans.* 194·7 cubic centimetres.

(6) A volume of gas standing in a bell glass over a mercury pneumatic trough measured 250 cubic centimetres. The barometer at the time stood at 754 millimetres, and the level of the mercury in the bell was found by measurement to be 65 millimetres above the surface of the mercury in the trough. Required to reduce the volume to the standard pressure of 760 millimetres. *Ans.* 226·7 millimetres.

(6) What would be the answer to the same problem had the trough been filled with water? *Ans.* 246·4 cubic centimetres.

(7) A closed vessel which displaces one litre of air is poised on a balance with weights whose volume is inconsiderable. The balance is in equilibrium when the barometer stands at 760 millimetres. If the barometer falls to 710 millimetres, how much weight will be required to restore the equilibrium and to which side must it be added? The temperature is assumed to be constant at 0°. *Ans.* Weight required, 85 milligrammes.

(8) Given the weight of one litre of dry air at 0° C. and 760 millimetres, as above, what will be the weight at 0° C. and 720 millimetres? *Ans.* It is obvious that the weight of

a measured volume of gas must be less in proportion as the total mass of gas (from which the measure is taken) expands. The formula $V : V' = H' : H$ applies to a limited volume of gas under a variable pressure. Considering now an invariable measured volume of such a mass of gas, it will be seen that $H : H' = W : W'$ ∴ $760 : 720 = 1\cdot 293 : x$.

Ex. 19. *Open Manometer.*—In answering the questions above the student will have learned that tension is the permanent internal elasticity of a gas, in virtue of which it resists external pressure. When the gas is free to expand, as when contained by a bag or a bell glass over a pneumatic trough, that resistance is the pressure of the atmosphere, more or less modified by the medium through which it is transmitted; and, according to Mariotte's law, the gas does expand or contract until an equilibrium is reached. When the gas is held in a tight vessel its volume is essentially invariable, and the tension is balanced by the resistance of the walls of the vessel against which it exerts, under ordinary circumstances, a great pressure. In practical chemistry it often becomes a problem of great importance to measure the tension of a mass of confined gas, and the instrument used for this purpose is called a manometer. One of the simplest of these the student should construct and keep for future use.

Select a stout glass tube about three millimetres internal diameter and 300 millimetres long; bend into the shape of a U, and bring the two arms as near together as practicable without choking the bend. Leave one of the arms straight open, but bend the upper end of the other at right angles and draw out to receive a rubber connector. Mount on a wooden stand (easily made by tacking together two pieces of thin board), fasten a millimetre scale between the tubes and fill the U with mercury through the open end until the level of the two columns stands midway of the height.

(1) If on connecting the manometer with a glass vessel the mercury stands in the straight arm 50 millimetres higher than in the other, what is the tension of the gas in the interior ? If it stands 50 millimetres lower, what is the tension ? Is any other observation required in order to express the tension numerically ?

Ex. 20. *Expansion of Air by Heat.*—Use a plain retort about fifty cubic centimetres capacity. Dip the open mouth in a pan of water, and clamp in position. Heat the body of the retort gently with a lamp. Allow to cool. Observe and explain all the phenomena.

(1) If the mouth of the retort were tightly corked would the heat produce any effect ? How could this effect be shown with the manometer ? Try the experiment, first drying the retort, tightly corking the mouth, passing a small glass tube through the cork, and uniting this tube with the manometer by a stout rubber connector. Try both heating and cooling the retort.

(2) In what two ways may the effect of heat on a mass of gas be manifested ? Show that these two effects must be proportional, and that one may be taken as a measure of the other.

Ex. 21. *Relation of Tension (or Volume) to Temperature. Law of Charles.*—Take a glass flask about 300 cubic centimetres capacity; tightly cork; through the cork pass two tubes, the first a small tube bent to connect with the manometer, the second somewhat larger (about four millimetres), which should only rise above the cork sufficiently to receive a rubber connector about 50 millimetres long, guarded by a pinch cock. Clamp the neck to a firm support, leaving the body of the flask free, so as to enable the experimenter to bring up under it a beaker sufficiently large to hold the flask with space enough for the movement of a stirring rod all round it. Begin by drawing through the flask (by means of a suction pump) air dried by passing through a chloride-of-calcium tube. Having next connected the manometer (but leaving the drying tube still connected with the vertical opening), pack the flask with broken ice. Wait for an equilibrium of temperature, and then close the pinch cock. Remove the ice (easily done by lowering the beaker, leaving the neck clamped), and replace it by ice-cold water. Keep stirring the water round the flask until the temperature has risen to 5°, a thermometer having been placed at one side, dipping under the water, so as best to facilitate the observation. Read now the difference in the height of the manometer columns and the height of the barometer. Proceed in the same way until the temperature has risen to 10°, and so, for every successive five degrees, or thereabouts, read the manometer, and also the barometer if the last is changing. When the temperature of the water bath nears that of the room apply a flame to the beaker, and thus continue the observations up to 70° or 80°. Leave the apparatus to cool for the next experiment. Make now a table giving in one column the temperatures and in a parallel column the corresponding tensions, and seek a relation between the values. They should correspond to the proportion

$$H : H' = 273 + t° : 273 + t'°$$

Then, since the volume of a mass of gas is, by Mariotte's law, proportional to its internal tension, we have also

$$V : V' = 273 + t° : 273 + t'°$$

(1) Obviously we should reach the same result by measuring the increased volume under a constant tension as by measuring, as here, the increased tension under a constant volume, but the last is the easier experimental problem. The proportions which express the law of Charles, simply mean that the air increases in tension or in volume $\frac{1}{273}$ of its tension or volume at 0°. It is not necessary however that 0° should be taken as the initial point of the actual experiment. It may be begun at any point, and the temperature raised or lowered at will, when the same relation will appear. Why is it important that the connecting tube between flask and manometer should be as small as possible.

(2) At what temperature would the tension of a confined mass of at 0° gas be doubled? At what temperature would it

theoretically become zero? or, assuming the gas to expand under constant pressure, at what temperature would the volume be increased one half?

(3) An open vessel is heated to 819°. What portion of the air which the vessel contained at 27° remains in it at this temperature?

(4) Obviously in comparing gas volumes or gas tensions we must have a standard of temperature, and 0° is usually assumed as that standard. Reduce the following volumes of gas measured at the temperatures and pressures annexed to 0° and 760 millimetres.

1. 140 c. c. $H = 570$ mm. $t° = 136°·5$ *Ans.* 70 c. c.
2. 320 c. c. $H = 950$ mm. $t° = 91°$ *Ans.* 300 c. c.
3. 480 c. c. $H = 380$ mm. $t° = 68°·25$ *Ans.* 192 c. c.

The following formulæ, easily deduced from the above proportions, will be useful in such reductions,

$$V' = V \cdot \frac{H}{H'} \cdot \frac{t'_0}{t_0} \text{ or } W' = W \frac{H'}{H} \cdot \frac{t_0}{t'_0}.$$

The first applies when the volume of a given mass of air varies under changes of temperature and pressure, the second when the weight of air displaced by a closed vessel, or contained in an open vessel, varies under the same circumstances. The term t_0 stands for the sum $(273 + t°)$, and is usually called the absolute temperature. From the several proportions under which the laws of Mariotte and of Charles may be expressed other formulæ may be deduced useful in special cases, which need not be considered here.

Ex. 22. *Air and Aqueous Vapour, Tension of Mixture.*—Use the same apparatus as left from the previous experiment, the flask assumed to be filled with dry air and united to manometer. Raise bath to 20°, and maintain steadily at that temperature. Open pinch cock until equilibrium is established, and then close, placing the cock as low down on the rubber tube as possible. Fill the tube above the cock with water, and while pinching tight the open mouth of the rubber tube relieve the cock, and, after squeezing the small volume of water into the flask, again shut. If all the water evaporates add a further portion in the same way, and so on. It will take some time before equilibrium is reached. Note then the increase of tension. Raise now the bath to 30°, 40°, and 50° successively, and at each temperature repeat the observation, adding more water as necessary. Remembering now that the tension of the dry air was measured by the height of the barometer at the time the pinch cock was closed, and that from this value the tension at any other temperature can be calculated, assuming also, according to a well-known principle, that when mixed with air the tension of aqueous vapour is added to that of the air, find the tension of aqueous vapour at each observation, and compare your result with that given in tables of the maximum tension of aqueous vapour.

(1) How can allowance be made for a change in the height of the barometer during the course of the experiment? If you measure the tension of a mass of air saturated with aqueous vapour at a given temperature, how can you find what would have been its tension under the same condition if dry?

(2) A volume of air standing in a graduated tube over a water pneumatic trough measures 75 cubic centimetres. The temperature is 20°, the height of barometer 770 millimetres, and the level of the water in the bell 270 millimetres above

that in the trough. What would have been the volume if $t = 0°$, $H = 760$ millimetres, and the air dry?

(3) As plainly indicated by the heading of this division, we have in the last six experiments studied atmospheric air as the type of aëriform matter and have tacitly assumed that all gases are affected equally by changes of pressure or temperature and conform sharply to the two laws we have described. This is not strictly true, but the details of the subject would be out of place in an elementary book and the differences are inappreciable in ordinary experiments. The composition and chemical relation of air will be considered later.

3. Oxygen Gas.

Ex. 23. *Preparation.*—Take eight grammes of pulverized potassic chlorate and two grammes of black oxide of manganese. Mix thoroughly in a mortar and introduce into an "ignition tube" of Bohemian glass. Close with a tight-fitting cork, through which passes a glass gas-delivery tube leading under the shelf of a pneumatic trough. The pneumatic trough may be cheaply made of sheet zinc by a tin smith, of such size as to hold one or two gallons of water, and the gas is best collected in ordinary glass fruit jars of a pint or a quart in capacity which can be hermetically closed with glass covers. The ignition tube must be mounted on some form of short-tube furnace, and the tube heater of the petroleum stove figured at the end of book serves an admirable purpose. Obviously, the cork must project in front of the furnace sufficiently to prevent scorching, and the powder must be gathered together in the heated portion of the tube. The heat should be gradually applied, and only raised to the highest temperature of the stove at the end of the experiment. This experiment is intended to give the student some experience with the methods of collecting and manipulating aëriform matter. He should from this point, if not before, mount his own apparatus, having been shown how to fit and perforate a cork and how to bend a glass tube.

For the mere preparation of oxygen gas no further details are required, but much more may be learned from the experiment.

In the first place, the production of a gas involves just as definite a loss of weight as would that of any other material. If several litres of oxygen escape from the ignition tube this tube will weigh less after the experiment by the exact weight of the aëriform material that has escaped. To test this let the student weigh the ignition tube before and after the experiment, and let him also measure the volume of the oxygen obtained. Under the ordinary conditions of a comfortably heated laboratory one litre of oxygen gas weighs about 1·34 gramme, and on this assumption it will be found that the total weight of the oxygen gas obtained nearly corresponds to the loss of weight observed. There are, however, obvious causes of error in addition to the rudeness of the tools here used which may seriously impair the sharpness of the result. 1. Under extreme conditions the above assumption may be materially in error, although if the thermometer and barometer are observed correction may now readily be made for this cause, knowing that under standard conditions a litre of dry oxygen weighs 1·4303 gramme. 2. The aqueous vapour in the gas collected over a water pneumatic trough will largely influence its volume, and at a rapidly increasing rate, as the temperature rises, and may produce a much greater effect than has been allowed in the estimate above given, although this, again, may be exactly calculated. 3. There is always more or less hygrosopic moisture in the materials used, which is expelled with the gas, and the error thus arising can only be avoided by a most careful preliminary drying. Still, as these errors in part compensate each other, the result is more satisfactory than could be expected.

In the second place, the relations of the materials in the combustion tube have undergone an essential change. To test this shake up the materials left in the tube with water, filter, and wash black residue. Dry and weigh on filter (the paper having been previously weighed). Evaporate the filtrate to dryness in a porcelain dish (tared), and weigh the saline residue. Compare these weights with the weights of oxide of manganese and of potassic chlorate originally taken, and draw your own conclusion as to the source of the oxygen gas.

In the third place, the final saline residue is not potassic chlorate. Test by tasting, and also by crystallizing a portion of the residue, and also some potassic chlorate, in two watch glasses side by side and compare form of crystals. Compare also black residue with the oxide of manganese used.

(1) What has happened to the oxide of manganese?
(2) What has become of the potassic chlorate? The white residue is called potassic chloride. What inference can you draw from your own observation as to the difference between potassic chloride and potassic chlorate?
(3) Reverse the calculation above. Taking the loss of weight of the tube as the weight of oxygen gas collected, divide this weight by the observed volume and deduce the density (the weight of one litre) under the conditions of the experiment.

Ex. 24. (*a*) *Combustion.*—Fill by displacement with oxygen gas a quick-sealing (pint) fruit jar. Dry the jar and dry the gas by passing through a chloride-of-calcium tube the current from one of the cylinders now familiar in commerce in which the gas is stored under pressure. Provide a deflagrating spoon (best made of sheet iron), so arranged that the bowl of the spoon, hung from a cross bar confined at the neck, will reach nearly to the bottom of the jar. Line the bowl with asbestos paper (previously ignited) and place on the paper less than half a gramme of red phosphorus. Holding the spoon just over the mouth of the jar, light the phosphorus, and with a quick but deliberate motion plunge the spoon into the jar and seal the mouth.* Notice that as the phosphorus burns away a white powder forms in the jar. After the glass is cold, open the mouth un-

---

* It would be still better first to put the spoon in place, and, after sealing the jar, light the phosphorus by a burning glass. This and all experiments involving the deflagration of phosphorus should be undertaken by inexperienced hands only under the most careful supervision. They are always attended with some risk, and burns from phosphorus are painful and difficult to heal. The ordinary fruit jar, unless specially annealed, will not usually bear the full intensity of the deflagration without breaking and sometimes flying in pieces. The glass can be partially annealed by placing the jar in a kettle of cold water, and, after bringing the water to boiling, removing the kettle from the stove and leaving it to cool; but it is also a safe precaution to use an insufficient amount of phosphorus to exhaust the oxygen. One pint will consume theoretically 0·63 gramme of phosphorus, and if less than half a gramme is used, as directed, the deflagration will not reach a greater intensity than the jar can usually stand. Another all-important precaution is to use red phosphorus, as directed above, which, although it does not ignite so readily as ordinary stick phosphorus, will burn very well if supported on asbestos paper. In all cases the teacher should himself carefully experiment until he has perfect command of his apparatus, and then he can judge whether this is a safe experiment for his students. In most cases he will undoubtedly think it best to confine all experiments with phosphorus to the lecture table. Common phosphorus should never be used in the laboratory, and even with red phosphorus the spoons with their covering when removed from the jars should be left under water, and after any residual phosphorus has been burned off they should be scrupulously cleaned before being put away. Moreover, the room should be carefully inspected after the class has left.

N. B.—In all experiments with fruit jars, in which it is important that the cover should hold gas tight, THE RUBBER WASHER MUST BE WELL GREASED.

der water, which, rushing in, will show that the gas has disappeared. This experiment is the most striking illustration we have that burning consists in the union of the combustible with oxygen. The sole product of the process is here a solid, and therefore visible.

(b) *Combustion in Air.*—Repeat the last experiment, following in every respect the same directions, but filling the jar with dry air instead of oxygen. The action is less violent, but the same product—called phosphoric oxide—is formed; hence burning in air is the same process as burning in oxygen. But on opening the jar under water we find that only one fifth of the volume of the air has been consumed in the process; hence four fifths of the volume must consist of a different substance, whose properties we shall study later.

(c) *Combustion of Carbon.*—Make a precisely similar experiment to (a), using a small lump of charcoal, weighing not over one gramme. In this case, although the coal in part disappears, there is no visible product. On opening the mouth of the jar it will be found to be still full of gas. This gas, however, is not oxygen. The properties are all changed. Dip a lighted match into the jar, and it is extinguished. It is so heavy that it can be poured like water from one vessel to another. Pour the contents upon some lime water—or, still

BURNING OF CHARCOAL. 59

better, baryta water—at the bottom of another jar, and shake the solution up with the gas. Compare with oxygen. This gas—called carbonic dioxide, often also carbonic-acid gas — is obviously the product of the burning; and this product was evidently formed by the union of carbon and oxygen.

(*d*) Take a smouldering slow match and dip the lighted end into a jar of oxygen; as soon as the match inflames remove and repeat the trial, thus following the level of the gas as it is consumed.

(1) Do the facts which you have observed in (*a*) justify the conclusion that the burning consisted in a union of phosphorus and oxygen? and that the white residue was the product of that union?

(2) Do the further facts observed in (*b*) justify the conclusion that one fifth of the volume of atmospheric air consists of oxygen?

(3) What becomes of the charcoal in (*c*)? and what must be the composition of the aëriform product left in the jar? Show that the facts observed support your inference.

(4) So far as your own observation has extended, what are the properties of oxygen gas? Is it the same material which was obtained as one of the products in the decomposition of water? How would you recognize the gas if you met with it as the product of a chemical process?

(5) What is the nature of combustion so far as illustrated by the above experiments?

4. Hydrogen Gas.

Ex. 25. *Preparation.*—Prepare hydrogen gas by pouring a mixture of one part of sulphuric acid and five parts of water over clippings of sheet zinc in a glass flask. Connect the glass flask by means of a cork and glass tube with a pneumatic trough, and collect the gas in glass fruit jars after the air has been driven from the flask. As each jar fills it may be lifted from the water and the gas burned at the mouth. For other experiments a self-regulating generator should be provided, from which the gas can be drawn as required.*

Ex. 26. *Density of Hydrogen.* — The great lightness of hydrogen gas can be illustrated by the teacher in various ways—as by filling a small balloon, or blowing soap-bubbles with it; also by decanting it from a large jar to a smaller, holding both jars with the mouths down. The student may test this very striking quality for himself by filling a jar with the gas by displacement, only

---

* Such a generator can easily be made from old glass bottles, and will be useful in several experiments. For the inside bell use a half-pint *tincture* of thin glass. Through the bottom of such a bottle bore several holes three to four millimetres in diameter, using as a drill the sharpened point of a three-cornered file dipped in turpentine or kerosene. Fill the bottle with zinc clippings, tightly cork, and pass through the cork a gas-evolution tube guarded by a pinch cock. Clamp this bell firmly by means of wooden stays fastened by twine, in a considerably larger, open, tumbler-shaped vessel, which can be made by cutting off (with a hot coal, "sprengkohlen") the neck of a common quart wine bottle, or a tall glass beaker may be used for the purpose. Lastly, fill the open vessel with dilute sulphuric acid. On opening the cock the acid will flow into the bell, and the gas generated by its action on the zinc will soon drive out the air. On then closing the cock the bell will fill with gas and drive back the acid water, when the chemical action will cease and the apparatus be left in a condition to yield a constant supply of hydrogen.

displacing the air from above downwards instead of the reverse, as in the case of oxygen. This is most readily done by resting the open mouth of the jar on a square of cardboard, supported by the ring of a retort stand, and passing the glass delivery tube from the generator through a hole in the cardboard (which it should tightly fit) to the very top of the jar. When the jar is full, and while the gas current is still flowing, the tube should be slowly withdrawn; and a little address is required to slip under the cap at the right moment, so as to prevent air from mixing with the hydrogen at the open mouth.

Hydrogen gas is about fourteen and a half times lighter than air, about sixteen times lighter than oxygen, and, under standard conditions, one litre of hydrogen weighs very closely 0·09 gramme.

Ex. 27. *Combustion of Hydrogen.*—The production of water by the burning of hydrogen should be shown on as large a scale as possible by the teacher, and the apparatus described in the author's New Chemistry is admirably adapted to this purpose. To observe the same effect, let the student replace the delivery tube of the flask used in Ex. 25 with a short piece of tube drawn out to a jet; and after making assurance doubly sure that all the air has been driven from the flask,*

---

* The explosion of "hydrogen flasks" is a very frequent accident in chemical laboratories. In such cases the injury is usually

light and burn the hydrogen at the jet. Hold now a cold and dry jar over the jet until the moisture which condenses collects in drops to a sufficient extent to render the nature of the product evident. Test in a similar way the products of a flame of common illuminating gas; and, after a perceptible amount of water has been formed, shake up in the jar some lime water, and prove by the resulting turbidity that carbonic dioxide has also been formed. Test likewise the products of the flame of a candle. In this connection the teacher should discuss at length the general features of the burning of hydrocarbon fuels, including both the nature of the products and the stages of the process.

Ex. 28. *Nature of Flame.*—Make the following experiments with the flame of a Bunsen lamp with the air valve shut, and also as far as practicable with the flame of a candle. 1. Press the flame down with a broken bit of glass, and notice that the shell, or mantle, only is luminous. 2. Adjust a piece of glass tube so as to draw off the combustible gas which forms the cone of the flame, and notice that it may be lighted at the upper end of the tube. 3. Press down on the

---

caused by the scattering of the glass; and the danger can be in great measure prevented by wrapping the flask in a towel before lighting the jet. For this experiment it is safer to use a hydrogen generator that has been tested.

flame a screen of fine-wire gauze, and notice that, although the combustible gas passes through, the burning mantle is cut off. 4. Press down on the flame the back of a cold iron spoon, and notice the soot which collects upon it. 5. Open the air valve of the Bunsen burner, and notice that when the flame loses its luminous power no soot is deposited. By means of these and similar experiments the teacher should enforce the following conclusions: 1. That flame is always burning gas. 2. That the action takes place in the outer mantle of the flame. 3. That a combustible will not take fire and continue burning unless its temperature is raised to the point of ignition and maintained at that temperature. 4. That in ordinary burning the required temperature is maintained by the heat developed from the union of the combustible with the oxygen of the air. 5. That when the burning gas of a flame is cooled below its point of ignition the flame goes out, and hence the efficacy of the wire gauze in the safety lamp. 6. That charcoal, not being volatile, burns without flame, but that here also, if the glowing coals are cooled below a red heat, the union with oxygen stops, and the fire is extinguished. In this connection the student should be shown the use of the mouth blowpipe, and the effects of the reducing and oxidizing flames should be explained.

(1) It is not expected that the student will understand the nature of the chemical process in Ex. 25. This will hereafter appear; but he should clearly recognize at this stage that the product is the same combustible gas obtained in the decomposition of water.

(2) Show that Ex. 27 confirms the analysis of water in Ex. 16.

(3) From what does the carbonic dioxide formed by a burning candle come? In what respects does the flame of a candle differ from the flame of hydrogen? What is the cause of the differences?

5. Sulphur.

Ex. 29. *Specific Characters.* — The student should be given a roll of brimstone and asked to study and describe its distinguishing properties, including colour, hardness, tenacity, specific gravity, fusibility, volatility, colour of vapour, and solubility in ordinary solvents.

Ex. 30. *Melting and Boiling Points.*—The melting point may be determined by heating a small bit of sulphur in a glass quill tube closed at the lower end by means of a bath of some liquid in which the tube is dipped. The bath generally used for determining melting points is a small beaker glass containing sulphuric acid heated over a lamp; but, as the use of hot sulphuric acid is not unattended with danger in the hands of the inexperienced, melted paraffine or castor oil had better be substituted in this experiment. The temperature of the bath at which the sulphur begins to melt is observed by means of a thermometer hanging so that its bulb dips in the same bath at the side of the tube; but the student will require some instruction and practice before he can obtain accurate results with this apparatus. To measure directly the boiling point of sulphur we require a peculiar thermometer made by Geissler, which is filled under pressure, and indicates temperature up to 450°. The sulphur is boiled in a small flask with a long neck, and the thermometer should be suspended so that its bulb hangs in the midst of the deep-red vapour. Such thermometers cost in Europe three dollars each, and unless provided this observation must be omitted.

Ex. 31. *Modifications of Sulphur.*—First, dissolve a gramme of sulphur in sulphide of carbon, and allow the solution to evaporate spontaneously, when the sulphur will crystallize. (Sulphide of carbon is very volatile and inflammable. HAVE NO LIGHTS NEAR BY.) Second, melt enough sulphur to nearly fill a small beaker, taking care that the temperature does not rise much above the melting point. Let the vessel cool until a crust begins to form, and then promptly pour out what remains liquid, which will leave the beaker lined with crystals of sulphur having a very different form and color from those first obtained. Third, melt some sulphur in a test tube, and raise the temperature until the liquid, at first very limpid, becomes thick and pasty, and then pour the material out in a fine stream into a basin of cold water. Let now the student study and describe, as well as he can, the differences between these three conditions of sulphur.

(1) Are the several modifications of sulphur the same substance or different substances?

Ex. 32. *Combustion of Sulphur.* — Burn a small amount of sulphur in a jar of oxygen gas, making the experiment as described in Ex. 24. The action is far less violent than in that experiment; and, in order to facilitate the burning, it is better to use shreds of asbestos soaked with sulphur (when melted) instead of a lump of brimstone. The asbestos is not acted on, and prevents the sulphur, when melted by the heat, from running together. When the jar is opened it will be found that it is still full of an aëriform material, but that the new product, although having the same volume, is wholly different from the oxygen gas with which the experiment began. The product is obviously a compound of sulphur and oxygen, and must weigh more than the initial oxygen by just the weight of sulphur burned. Indeed, as will hereafter appear, it weighs just twice as much as the same volume of oxygen. Again, unlike oxygen, it has a very suffocating odor, and by this will be recognized as the same product which is so offensive from a burning match. Let the student immerse in the gas a red rose, or some other highly coloured flower, and witness the remarkable bleaching power. Let him then dissolve the rest of the gas by shaking up 50 cubic centimetres of water in the jar. The gas is called sulphurous oxide, and the solution in water, which has acid qualities, is well known as sulphurous acid. Taste the solution. Dip into it, momentarily, a strip of paper coloured blue with litmus.

(1) Justify the inference that sulphurous oxide is composed of sulphur and oxygen.

Ex. 33. *Production of Sulphuric Oxide.* — Take a short length of small combustion tubing (100 millimetres long and from 6 to 8 millimetres bore), fill, but not too tightly, the middle portion of the tube with platinized asbestos,* tightly cork one end and draw out the other to a small tubulature; through a perforation in the cork pass a small glass tube. Wire the combustion tube in a horizontal position to the ring of a retort stand. Take also a test tube corked tight-

---

* Platinized asbestos can be purchased of dealers in chemicals but is easily made by drenching asbestos wool with solution of platinum chloride and then igniting. Asbestos paper can be treated in this way, and affords a convenient preparation, since it can be rolled into a shape that will just fit and fill the tube.

ly; make two perforations in the cork. Through one of these pass an inlet tube extending to the bottom of the test tube, and through the other an outlet tube only extending through the cork. Stand the test tube in a beaker and pack round it broken ice and connect the inlet tube by a rubber connector with the tubulature of the combustion tube. Connect the other end of the combustion tube with a bent glass tube that will reach to the bottom of a fruit jar. Use a quart fruit jar, and, having filled it by displacement with *dry* oxygen—Ex. 24 (*a*)—repeat Ex. 32, using only half a gramme of sulphur. This amount is not sufficient to exhaust the oxygen, so that at the end of the combustion there must remain in the jar a mixture of sulphurous oxide and oxygen gases. After the jar has cooled uncover, remove the deflagrating spoon, and adjust the bent tube so that it leads to the bottom of the jar, as above described. Place a Bunsen lamp under the combustion tube and heat the platinized asbestos to dull-red heat. Connect, lastly, the other end of the apparatus with an aspirator, and slowly draw the contents of the jar over the asbestos and through the test tube. Notice that during the process the platinized asbestos undergoes no change whatever, but there will collect in the test tube a considerable amount of a white crystalline solid.

After dismounting the apparatus dissolve the

SULPHURIC ACID. 69

substance collected in the cold test tube in a few cubic centimetres of water, and compare the properties of this solution with those of the dilute sulphuric acid used in Ex. 25. For this purpose prepare a very weak solution of baric chloride and half fill with it two test tubes standing side by side. Add to each a few drops of hydrochloric acid. Add to the first a few drops of what is known to be diluted sulphuric acid, such as used in Ex. 25; add to the second the product of the last experiment, and compare results. Repeat now Ex. 24 (*a*), and dissolve the white product of that combustion in a few cubic centimetres of water; notice the time taken for complete solution. Taste, test with litmus paper, and also with baric chloride in the same way as above. Obviously, then, both phosphorus and sulphur, by uniting with oxygen in the process of combustion, yield compounds which, when dissolved in water, give products that have a strong acid taste.

Phosphorus, oxygen, and water yield phosphoric acid.
Sulphur, oxygen, and water " sulphurous "
" more oxygen, and water " sulphuric "

How far these cases are illustrations of a general principle will appear hereafter.

(1) Are we justified in drawing the conclusion from this experiment that sulphuric oxide differs from sulphurous oxide and sulphuric acid from sulphurous acid only in holding more oxygen in combination?

(2) On the basis of the facts hitherto observed, is the evidence above given (that the final product of this experiment is identical with common sulphuric acid) satisfactory?

6. Chlorine.

Ex. 34. *Preparation of Hydrochloric-Acid Gas.*—Mix 20 grammes of sulphuric acid with 4 grammes of water, pouring the acid slowly into the water, and when cold add this mixture to 10 grammes of powdered common salt in a glass flask. Cork the flask and provide a glass tube passing through the cork and leading to the bottom of a fruit jar. Cover the mouth of the jar with a card (Ex. 26), and collect the heavy, suffocating gas by displacement. Three full jars will be required for further experiments. These, before filling, must be thoroughly dried, and the dense fumes which form when hydrochloric-acid gas mixes with the air will always show when a jar is filled to overflowing. To one of the jars prepared as above add 10 grammes of water, which will instantly absorb the contents; and the solution thus obtained, although weaker, is the same preparation as the liquid hydrochloric acid so much used in chemical laboratories. The commercial acid often contains over four hundred times its volume of dissolved gas.

Ex. 35. *Composition of Hydrochloric-Acid Gas.*—To another jar of the gas obtained in the last experiment add 20 or 30 grammes of sodium amalgam, and, instantly sealing the jar, shake up the amalgam with the gas as long as absorption continues. Then open the jar under water,* which will rush in and show that one half of the volume has been absorbed. Apply a lighted match to the residual gas, and it will be found to be hydrogen. The only other ingredient of hydrochloric-acid gas will appear in the next experiment.

Ex. 36. *Preparation of Chlorine Gas.*—Fill a flask (50 cubic centimetres capacity), fitted with a perforated rubber stopper and outlet tube, to three fourths of its capacity with lumps of black oxide of manganese. Pour upon these lumps strong liquid hydrochloric acid so as to fill the interstices only. Allow the flask to stand for a short time, and then apply a gentle heat. A yellowish gas comes off in abundance, which is much heavier than the air, and can be collected by displacement if the outlet tube reaches quite to the bottom of the jars used for the purpose, and the mouth is covered with a disk of cardboard, as already described. Chlorine gas is very suffocating, and the smallest puff, if inhaled, may produce serious results. This experiment should therefore be performed with extreme caution,

---

* Use a glass or porcelain dish to hold the water. Not the pneumatic trough, which would be corroded by the mercury falling into it.

either in the open air or under a hood with a strong draught. The colour of the gas shows when a jar is full, and three jars thus filled are required for the following experiments: 1. In the first of these jars plunge some tinsel or other metal leaf hanging from the end of a long stick, and almost every metal will at once enter into direct union with the chlorine, often with ignition. 2. Place mouth to mouth a jar of hydrogen over a jar of chlorine, and, holding the open mouths together confined by a rubber band, invert the two, and in a few moments, when the gases have mixed, loosely cover (but on no account seal) both of the jars. If now one of the jars is exposed to bright daylight (not direct sunlight), a gradual union between the chlorine and the hydrogen gases will take place, and after the yellow colour has disappeared the product can easily be recognized as hydrochloric acid gas. The second jar, kept in the dark, will undergo no change, and can be used for comparison. If now this jar is exposed to direct sunlight, the same combination will suddenly take place with explosive violence.* This experiment is a dangerous one, and should only be made with the greatest caution, the mouth of the jar being

---

* There is more or less danger in all stages of this experiment, not only from the violence of the explosion, but also from the risk of breathing a puff of chlorine. It should never be intrusted to careless hands, nor indeed to the hands of any student before all the necessary precautions have been pointed out and enforced.

loosely covered with a pasteboard disk. The composition of hydrochloric acid is thus fully established. 3. Into the third jar of chlorine, prepared as above, pour 200 cubic centimetres of water, and, after closing the jar, shake the water up with the gas, which will be almost completely absorbed and impart its colour to the solution. Soak then in the water a strip of calico printed with madder, and notice how rapidly the colour is discharged. In this connection the use of chlorine as a bleaching agent should be explained.

(1) What proof has been given that hydrochloric acid gas consists solely of hydrogen and chlorine? Do our experiments show in what proportions hydrogen gas and chlorine gas combine by volume?

(2) What is the composition of liquid hydrochloric acid? Have you noticed any difference between the union of hydrochloric-acid gas with water and ordinary solution? Does the composition of liquid hydrochloric acid conform to the general scheme exhibited under Ex. 33? Point out agreements and differences.

(3) Compare sulphuric acid and liquid hydrochloric acid, *using the ordinary laboratory acids diluted with four or five times their volume of water.* Try taste, litmus paper, and action on zinc clippings. Also evaporate a few drops of each on watch glasses and observe effects.

7. Carbon.

Ex. 37. (*a*) *Preparation of Charcoal.*—Take small billets of three or four different kinds of wood, including the densest and lightest that can be procured. Cover with sand in an iron crucible, and heat to redness until the smoking stops. It is best to light the gas thus given off above the sand to prevent it from escaping into the room.

(*b*) Heat to redness over a lamp in a porcelain crucible some lumps of sugar as long as any vapor is evolved. In this connection the general facts in regard to the composition of organic substances should be briefly stated, and the relations of carbon as the non-volatile skeleton of organized matter should be explained. Also, the relations of diamond and graphite to charcoal should be stated, and compared with those between the different states of sulphur.

Ex. 38. *Specific Characters of Charcoal.*—The student should be asked to study the distinguishing characters of the charcoal prepared in the last experiment, and to compare these with those of sulphur already studied. If the work is judiciously directed and criticised this exercise will be very instructive; and, for the very reason that the two substances are so different, it will be a good preparation for comparing hereafter two substances which are closely alike. As charcoal is a porous body whose external volume depends on that of the organized material from which it was made, the density must obviously be left out of consideration in this comparison.

Ex. 39. (*a*) *Preparation of Carbonic Dioxide* (*Carbonic-Acid Gas*).—This gas, as we have seen, is formed by the burning of charcoal (Ex. 24, *c*); it is also easily made by the action of liquid hydrochloric acid on marble (calcic carbonate). Half fill a glass flask (250 cubic centimetres capacity) with small lumps of marble. Pour upon the marble common muriatic acid (the commercial name of crude liquid hydrochloric acid) mixed with three times its volume of water. Connect with a pneumatic trough and collect in the usual way. Half fill one of the jars, and, after closing, shake the gas up with the remaining water. Open from time to time to admit air until the absorption ceases. At the ordinary pressure of the air water will absorb its own volume of carbonic-dioxide gas, and soda water is the same solution under pressure. This aqueous solution may be called carbonic acid. Dip into it a strip of blue litmus paper and notice the difference between the effect of this weak volatile acid and that of a strong fixed acid like sulphuric acid when both are equally dilute.

(*b*) *What was the Source of Carbonic Dioxide in Last Experiment?*—To answer this question, prepare some lime water from quicklime slaked as in Ex. 15. Fill a quart fruit jar somewhat over one half (four sevenths) with lime water and the rest with carbonic dioxide (pouring in the heavy gas just as you would a liquid). Close, and shake the gas and water well together, admitting air from time to time as the absorption goes on. Allow the precipitate to settle; pour off the clear water, collect the precipitate on a filter, wash, and dry. Transfer next the powder to a test tube and pour on a few cubic centimetres of dilute hydrochloric acid. Notice the effervescence and recognize as carbonic dioxide the gas evolved. To make the demonstration complete the student must be told that marble is one of the many mineral forms of carbonate of lime and is chemically the same substance as the white powder thus prepared. The proof of this will appear later. In this connection the important relations of carbonic dioxide in nature should be discussed, its presence in the breath should be shown, and its association with alcohol as a product of fermentation and its presence in beer, sparkling wine, and other effervescing drinks, should be explained.

(c) Repeat Ex. 24 (c), and after the jar has cooled open the mouth under water. There will be no expansion of the aëriform product, and no contraction except that due to the slow solution of the gas in water. Obviously, then, a given volume of oxygen gas yields the same volume of carbonic-dioxide gas, and the last must weigh more than the first by the weight of the charcoal which the oxygen gas absorbs in the process of burning,

(1) Is the volume of the atmosphere altered by the smoke which our fires pour into it?

Ex. 40. (a) *Production of Carbonic Oxide.*— Provide two rubber gas bags, holding about one litre each. Connect these with the ends of a length of hard Bohemian glass tubing, which should be filled, but not tightly packed, with finely pulverized charcoal that has been thoroughly burned. Having filled one of the bags to about one half of its capacity with carbonic dioxide, heat the tube over two or more Bunsen lamps (best a gas tube furnace) to a full red heat,* and pass the gas slowly backwards and forwards so long as any increase of volume is perceptible, and at last it will be found that the volume has doubled. Remove now the full bag and transfer a portion of the product to a small glass jar over the pneumatic trough. Lift the jar from the water with the mouth down and the gas will not at once escape, because the new product is even lighter than air. It may now be lighted at the mouth of the jar and the peculiar color of the flame noticed and the product of the combustion shown to be carbonic dioxide. In this experiment it is evident that the carbonic dioxide must

---

* The kerosene stove does not give a sufficiently high temperature. By burning alcohol in the stove, however, the requisite heat can be obtained; but care should be taken to avoid explosions, and it will be safer to reserve this experiment for the lecture table.

have united with the material of the charcoal, and the new product, called carbonic oxide, must differ from carbonic dioxide only in containing more carbon or, what amounts to the same thing, proportionally less oxygen. The same inference may be drawn from the fact that in burning carbonic oxide changes back to carbonic dioxide.

(*b*) Add ten grammes of oxalic acid to a small flask, corked and fitted with an evolution tube leading to a pneumatic trough. Pour over this crystalline solid five or six times its weight of strong sulphuric acid (oil of vitriol). Support (on a retort stand) the flask, protected by a square of asbestos paper, and apply gentle heat. The gas which comes over copiously is a mixture of carbonic oxide and carbonic dioxide. The last will be slowly absorbed by the water, and very rapidly if some caustic soda is added to the pneumatic trough. Collect in fruit jar, and, after allowing to stand until absorption is ended, compare this product with that of last experiment.

Ex. 41. *Ethylene* (*Olefiant Gas*).—Pour into a fifty-cubic-centimetre flask five cubic centimetres of high-proof alcohol, and then add slowly twenty cubic centimetres of strong sulphuric acid. Connect with a pneumatic trough and heat carefully, protecting the glass by interposing asbestos paper between the flask and the lamp. Ethylene, the gas thus obtained, burns with a brilliant flame and is one of the constituents of illuminating gas. The sole products of its combustion are carbonic dioxide and water, and it must therefore contain both carbon and hydrogen. It is here selected as an example of a very large class of substances called hydrocarbons. The phenomena attending their combustion have already been discussed (Ex. 28). In illuminating gas, a very complex product, ethylene is mixed, among other things, with a very large amount of another hydrocarbon, called methan (or marsh gas), which contains only half as much carbon and has far less illuminating power. The petroleums and the products obtained from them, known as benzine, kerosene, astral oil, paraffine, etc., are chiefly mixtures of a great number of hydrocarbons (gases, liquids, and solids), resembling in their chemical relations marsh gas, and classed with it under the general designation of the paraffines. Olefiant gas prepared as above is so called because it unites directly either with chlorine or bromine to form a liquid which has an oily aspect, and there are several other hydrocarbons formed in the distillation of coal which resemble olefiant gas in this respect and are classed with it under the name of the olefines. Then there is a hydrocarbon containing only one half as much hydrogen as olefiant gas, which is formed abundantly when a Bunsen lamp burns at the base, and is at once recognized by its unpleasant odour. This hydrocarbon is also one of a class known as the acetylenes, and is itself usually called by the same name. Lastly, there is a very remarkable class of hydrocarbons obtained by the distillation of coal tar, of which benzol and toluol are the chief members and from which the aniline dyes are produced. These three classes, although the most important groups, by no means include all the known hydrocarbon compounds, while the possibilities of multiplication are unlimited, and from the great family of hydrocarbons the almost endless products of organic chemistry may be derived. The points here suggested the teacher will expand as he sees fit.

(1) Charcoal graphite or diamond when burnt in oxygen gas all yield the same product (carbonic dioxide). Are they the same substance ?

(2) Does a solution of carbonic acid in water conform to your general conception of an acid ?

(3) Compare the composition of carbonic oxide and carbonic dioxide with that of sulphurous oxide and sulphuric oxide. Do all these oxides form acids by uniting with water ? Regarding the intensity of the acid taste and of the acid reaction as some measure of the strength of the acids, what would be your estimate of the relative strength of the acids thus formed, and how does this degree of strength compare with the apparent attraction of the oxides for water ?

(4) Is the proof here given that olefiant gas is composed of carbon and hydrogen satisfactory ? Is it equally clear that the gas consists only of carbon and hydrogen ?

8. Nitrogen.

Ex. 42. *Preparation of Nitrogen Gas.*—The aëriform product left in the jar from Ex. 24 (*b*) is nitrogen gas. The student should test the gas by immersing in it a lighted match; but the elementary student can not be expected to learn through actual experiments the complex relations of this remarkable substance. It should, however, be made evident by the teacher that the inability to support combustion is in entire harmony with the general inert relations of nitrogen gas, and that this is its most striking characteristics. Nevertheless, when the necessary conditions are fulfilled, nitrogen readily enters into combination, especially with oxygen and hydrogen, forming a numerous and important class of products. In illustration of this last point the formation of nitre should be explained; and, starting from this natural product, the student may prepare a few well-marked nitrogen compounds as follows :

Ex. 43. (*a*) *Preparation of Nitric Acid.*—Mix in the body of a small glass retort 30 grammes of nitre with the same weight of sulphuric acid. Allow the mixture to stand for several hours, and then distil over 15 grammes of a yellowish liquid, which is nitric acid (a very important chemical agent, consisting of nitrogen combined with both oxygen and hydrogen). The yellow colour of the product is due to an admixture of another nitrogenized product, which is very volatile, and may be driven off by gently heating the acid in a flask.

(b) *Nitric Acid contains Oxygen.*—Take in a test tube four or five cubic centimetres of the acid just made. Drop into it in small portions at a time not over one decigramme of coarsely pulverized roll brimstone.* Cautiously heat to boiling and maintain gentle ebullition for some time. Largely dilute with water and decant into a clean test tube from the remaining sulphur. Test with solution of baric chloride as described in Ex. 33. It will thus appear that sulphuric acid is formed by action of nitric acid on sulphur. Remembering now that sulphuric acid consists of sulphur, oxygen, and water, draw your own inferences.

Indeed, nitric acid contains so much oxygen, held feebly in combination, that it furnishes an efficient means of uniting oxygen to other bodies. It may sustain combustion like the atmosphere, and it is for these reasons said to be a powerful oxidizing agent. The teacher can effectively illustrate this point by pouring from a long-handled glass or porcelain spoon a few cubic centimetres of the strongest acid on finely pulverized charcoal. The powder should be well dried by heating it to

---

* The action of very strong nitric acid on combustible matter is often violent, and this experiment should be made with caution.

COMPOSITION OF NITRIC ACID. 83

incipient redness in a porcelain dish, and while still hot the acid poured upon it (a few drops at a time). The charcoal will then flash almost like gunpowder. In this connection the teacher may add that in the explosion of gunpowder charcoal burns at the expense of the oxygen stored in the nitre.

(*c*) *Nitric Acid contains Nitrogen and Water.* —This can be shown by passing the vapor of the acid carried by a current of carbonic dioxide over copper clippings heated to redness in a combustion tube. The metal will take up all the oxygen in the acid except that belonging to the water present, while the water set free may be collected by passing the current after leaving the combustion tube through a U-tube packed in ice. If, lastly, the current is passed on to a small pneumatic trough filled with water holding caustic alkali in solution the carbonic dioxide will be absorbed and nitrogen gas collected. A regulated current of carbonic dioxide is easily obtained with the generator described in note to Ex. 25, filling the bell with broken marble and the beaker with common muriatic acid diluted with an equal volume of water. The few cubic centimetres of nitric acid required are best held in a bulb tube so supported that it may be warmed with a lamp, and connected at one end with the generator and at the other with the combustion tube by a rubber connector (corks would be instantly corroded). The combustion tube may be arranged as in Ex. 33, but should be somewhat larger, and is best filled with finely pulverized copper, such as is obtained by the reduction of copper oxide. The small pneumatic trough is easily extemporized out of a glass or porcelain dish and a large test tube. Other details of the apparatus may now be left to the ingenuity of the student; but the experiment should not be intrusted except to skilful manipulators, and in most cases will best be shown on the lecture table. In all cases, however, he should make a sketch of the apparatus in his note book and point out the use of each part and justify the conclusion that

Nitrogen, oxygen, and water yield nitric acid.

(1) Is the proof which has been given of the composition of nitric acid synthetical or analytical? The synthesis of nitric acid can not be readily made because, in conformity to its great inertness, nitrogen does not combine directly with oxygen. Nevertheless, by indirect means, the oxide of nitrogen corresponding to nitric acid has been prepared. It is a white crystalline solid which eagerly unites with water. Compare sulphuric oxide.

(2) You have now handled the most important acids used in a chemical laboratory. Make a list of them with the composition of each so far as you have discovered it. Make clear that you can recognize all of them whether concentrated or diluted. In every case a specific test has not been given, but by inquiry of your teacher or elsewhere seek the necessary knowledge until you are perfectly certain of your ability to distinguish all these substances.

Ex. 41. (a) *Preparation of Nitric Oxide.*—Place fifty grammes of copper clippings in a glass flask fitted with a cork, through which passes a tube funnel as well as an evolution tube. Drench the clippings with water, and then pour on through the funnel in successive portions the product of the last experiment mixed with three times its volume of water. After the action has started add a fresh portion from time to time as the effervescence slackens. Connect with a pneumatic trough and collect the gas in a quick-sealing jar. The colorless gas is a compound of nitrogen and oxygen, called nitric oxide. The deep-red fumes which appear in the generating flask, and whenever the nitric oxide mixes with the oxygen gas of the air, is another compound of nitrogen and oxygen containing more oxygen and called nitric peroxide; and it is chiefly this adventitious product which imparts the yellow tint to the crude nitric acid.

(b) *Analysis of Nitric Oxide.*—In a pint jar of nitric oxide collected in the last experiment burn a small bit of phosphorus, not exceeding one fourth of a gramme in weight, with all the precautions stated in Ex. 24. Notice that the same white product is formed as when phosphorus burns in pure oxygen gas or in air. Hence we may infer that nitric oxide contains oxygen. After the jar is cold open the mouth under water.

Notice that the residual gas fills only one half of the volume of the jar, and, further, that it has the characteristic inertness of nitrogen. Hence the additional conclusion that nitric oxide consists of equal volumes of nitrogen gas and oxygen gas.

(1) Compare the action of copper on nitric acid in the last two experiments. Notice that while in the first it reduces the acid to water and nitrogen gas, in the second it does not remove so much oxygen and leaves nitric oxide. Observe also that nitric oxide shows no tendency to unite with water.

Ex. 45. *Preparation of Ammonia.*—Mix in a gasometer* one volume of nitric oxide with two and a half volumes of hydrogen, and pass a slow

---

* A very useful and inexpensive gasometer may be made of a large glass bottle of the capacity of two quarts or one gallon, as required. Fit tightly in the neck a rubber cork with three perforations. Through these perforations pass glass tubes, all bent at right angles a short distance above the cork. Two of the tubes should reach the bottom of the bottle, the third only pass through the stopper. Connect one of the longer tubes with a sink by a length of rubber hose. In the same way connect the second of the longer tubes with a water tap; and, lastly, slip on to the shorter tube a third length of rubber hose to serve as an outlet for the gas. Guard all three of the tubes with pressure taps. Begin by filling the gasometer with water, allowing the liquid to flow in from the tap and the air to escape from the outlet. When full, close the outlet, open the overflow into the sink, remove the rubber hose from the water tap, and connect it with the gas generator. The gas will then flow in, the displaced water running off by the overflow. When a sufficient amount of gas has been collected the overflow must be closed and the rubber hose removed from the generator and replaced on the water tap. Then, on opening the tap, water will flow in again and drive out the gas by the outlet tube through any apparatus with which it may be connected.

stream of this mixture through a short tube filled with platinized asbestos and heated to a low red heat. There will be an abundant formation of aqueous vapours, indicating that a combination has taken place between the hydrogen gas and the oxygen of the nitric oxide; and at the same time there will be developed a strong pungent odour, familiar to every one as the odour of ammonia, which must evidently be formed by the union of nitrogen with hydrogen. This pungent product is a gas, and common aqua ammonia is a solution of the gas in water. To obtain a more familiar acquaintance with this important chemical agent, half fill a small glass flask with concentrated aqua ammonia and close the neck with a rubber stopper, through which passes an evolution tube leading to the top of a quick-sealing glass jar, arranged as in Ex. 26. Heat the liquid in the flask to boiling, when a large volume of colourless gas will pass over and may be collected by displacement. When the jar is full, the overflow will be at once recognized by the strong odour. The process may then be stopped, the jar closed, and the gas preserved for another experiment.

Ex. 46. (*a*) *Ammonia Salts.*—Mix five cubic centimetres of strong aqua ammonia with twice its volume of water. In the same way mix five cubic centimetres of strong nitric acid with twice its volume of water. Study the effects of these solutions on test papers, both litmus and turmeric. Next add slowly the ammonia to the nitric acid until the opposite effects exactly neutralize each other. Lastly, evaporate the mixture at a low heat until a drop taken out on a rod solidifies, and then when the dish is set on one side a white salt, called ammonic nitrate, will crystallize out.

(b) Make the same experiment with sulphuric acid, with phosphoric acid, with hydrochloric acid, and with carbonic acid. In the last case add the fifteen cubic centimetres of diluted aqua ammonia to a jar of carbonic-acid gas and shake together. Evaporate each solution to dryness and collect the crystalline salt.

Obviously the solution of ammonia gas (aqua ammonia) sustains relations which are the very opposite to those of the acids, and belongs to an equally important class of chemical products variously called alkalis (when the solutions are caustic) or bases (when they are not). The acids exhibit no tendency to unite with each other, but they eagerly unite with the *anti-acids* (as in the above experiments) to form a class of bodies, for the most part crystalline, called salts. The anti-acids are formed, as a rule, by the union of metals with oxygen and water, while the acids (as we have seen) result as generally from a similar union of bodies, like phosphorus, sulphur, and carbon, which do not exhibit metallic characters. Having studied the relations of a few of these so-called metalloids we pass next to study the relations of several metals.

9. Magnesium and Zinc.

Ex. 47. *Specific Characters.* — The student should be given a gramme of magnesium ribbon, and with this he should study the characters of the metal (colour, lustre, tenacity, specific gravity) and compare these with the corresponding properties of metallic zinc, also given to him rolled out into ribbon of about the same size. Reserving a short length of the magnesium ribbon for burning, the student should dissolve the rest in a few cubic centimetres of dilute sulphuric acid. Collect and examine the gas evolved, and compare the reaction with that in Ex. 25. Lastly, let him evaporate the solution of magnesium thus obtained on a watch glass, and compare the crystalline residue with that obtained from the solution of zinc formed in the above-cited experiment.

Ex. 48. *Burning of Magnesium.* — Let the student burn the reserved piece of magnesium ribbon, holding it by pincers and lighting it like a match, and let him compare the combustibility and colour of the flame with that produced with a similar ribbon of zinc. In order that he may further study the nature of the product formed, furnish the student with half a gramme of magnesium* powder. Let him place this on a small square of asbestos paper (previously ignited) and weigh the amount accurately on the pan of a balance. Let him now ignite the powder with a match, and when burned out and cooled reweigh it. What means the increase of weight, and what must be the composition of the white powder left? Transfer the powder to a small evaporating dish. Thoroughly drench with water. Place a small bit of the wet powder on red litmus paper. Compare the effect with that of an acid. Lastly, dissolve the residue in the smallest possible amount of dilute sulphuric acid, adding the acid drop by drop. Evaporate the solution till a crust appears and leave to crystallize. Can you recognize the saline product by the taste?

Try the same experiments with zinc; but its powder does not burn so readily, and it is more difficult to recognize the increase of weight. Nevertheless, it is easily burned by sifting the powder on to a sheet of paper through the flame of a Bunsen burner held obliquely, or by spreading

---

* A few years ago magnesium would have been too expensive for general experimenting, but, as a result of its application in the arts, it can now be purchased for a moderate price. At the price quoted by German dealers in chemicals the two grammes required for each student would cost less than three cents.

the powder over a square of asbestos paper and playing on it with the same flame.

(1) The comparison of two metals closely resembling each other, like magnesium and zinc, affords excellent practice, and may be used to test the student's skill in observation and deduction. If further practice is thought necessary a comparison may be made between two metals resembling each other still more closely; as, for example, iron and nickel. In all such cases the student should be required to work out the results unaided, and make a full and clear statement in his note book of what he observes and what he infers. His work should then be carefully criticised, and, if necessary, the experiments repeated, after fresh directions or suggestions from the teacher. The student should be led to appreciate the fact that although the distinctions between substances are usually broad and clear, they are also at times narrow and indefinite, and that the identification or differentiation of a newly found substance often turns on minute observations and delicate discriminations.

(2) How can you prove that magnesic oxide combines with water? Does zinc oxide combine in like manner?

Compare—

Magnesium, oxygen, and water yield magnesic hydrate —basic.

Sulphur, oxygen, and water yield sulphuric acid—acid.

Further, it appears that—

Magnesic hydrate and sulphuric acid yield Epsom salts —salt.

Would magnesic oxide yield the same product as magnesic hydrate?

(3) Compare the product of the action of magnesium on dilute sulphuric acid with that obtained by dissolving magnesic oxide in the same solvent. Why is hydrogen gas not evolved in the second process?

10. Sodium.

Ex. 49. *Specific Characters.*—The specific characters of this interesting alkaline metal should be shown as far as possible to the student; but it will seldom be advisable to intrust the material to inexperienced hands, and equally good practice in studying specific characters can be had with cheaper and less dangerous substances. The action of sodium on water illustrates principles so fundamental in the theory of chemistry that the experiment should on no account be omitted. The action of the pure metal is always violent, and frequently dangerous; still, it is a very interesting experiment, which the teacher may make before the class, with proper precautions. The best way is to throw a bit of sodium, not larger than a pea, on some sheets of porous paper thoroughly soaked and running with water. The melted globule is thus prevented from swimming round, and the heat developed by the chemical change accumulates to such an extent as to inflame the escaping hydrogen, which burns with a flame that is coloured yellow by the presence of sodium vapour. For use of students, an amalgam of sodium—one part of sodium to about ten parts of mercury—should be prepared; and this may be used with entire safety. The action of sodium on mercury is violent, but the amalgam can be easily made by heating the mercury to about 200° in a Hessian crucible of eight or ten times the capacity required to hold the metal, and then adding the sodium in one large bar. Assuming this amalgam to have been previously prepared or purchased, the student should make the following experiment, having, of course, been previously told the object of using the mercury, and that it plays no part in the chemical change : Take a small flask fitted for the evolution of gas, and place in it about twenty-five grammes of water ; add now a few lumps of the amalgam, and collect the gas over the pneumatic trough. As the evolution lessens, hasten it by heating the flask with a lamp. Burn the gas that is collected, and recognize that it is hydrogen. Pour off now the solution left in the flask from the mercury, and, in the first place, test it with litmus and turmeric paper which have previously been dipped in a very weak acid. It will thus be seen that the product is a substance which, like the solution of ammonia, reverses the effect of an acid on vegetable dyes ; in other words, that it is basic. Such soluble and caustic bases are called alkalies. Rub a few drops of the liquid between the fingers, and notice the effect, which is termed caustic. Evaporate now the liquid, and compare the residue with caustic soda. Redissolve this residue in a very small amount of water, and divide the solution between three watch glasses; neutralize the solution in the first glass with a few drops of hydrochloric acid, that in the second glass with a few drops of nitric acid, and add the contents of the third watch glass to about twenty-five cubic centimetres of carbonic acid (soda water). Allow the solutions to evaporate, and examine the crystals formed with a lens; also attempt to recognize the products by tasting the residue in each case; they will be discovered to be common salt, sodic nitrate, and sodic carbonate, respectively. As caustic soda has thus been made solely with sodium and water, a probable inference in regard to its composition may at once be drawn, and the three familiar products last obtained will be recognized as salts of sodium. In this connection the student should be told about the sources of these substances, and their uses in daily life and in the arts.

Ex. 50. (*a*) Using a small amount of magnesium or zinc powder and boiling with water in a test tube, compare the action of these metals on water with that of sodium. It will appear that—

Sodium acts violently on water and yields sodic hydrate and hydrogen gas.

Magnesium acts slowly on water and yields magnesic hydrate and hydrogen gas.

Zinc acts very feebly on water and yields zinc oxide and hydrogen gas.

(*b*) The teacher should burn sodium in dry

oxygen, first melting the metal in an iron spoon. Dissolve the oxide formed in water, evaporate to dryness, and compare it with product of the direct action of sodium on water. Care must be taken to separate the white powder from unburnt metal before adding water, and the experiment should not be intrusted to unskilful students. It will now further appear that—

Sodium unites with oxygen to form sodic oxide (white powder).

Magnesium unites with oxygen to form magnesic oxide (white powder).

Zinc unites with oxygen to form zinc oxide (white powder).

Also that—

Sodium oxide unites with water eagerly to form sodic hydrate.

Magnesium oxide unites with water feebly to form magnesic hydrate.

Zinc oxide will not unite with water.

The striking differences depend on the fact that sodic hydrate is very soluble in water. Interpret all the phenomena.

(c) The teacher should burn sodium in dry chlorine. It burns readily when heated above melting point in an iron spoon, but the experiment should be made under a hood with powerful draught. A considerable part of the product volatilizes. This is readily dissolved in water and crystallized, when the cubic form of the crystal and taste show it to be common salt. It thus appears that sodium and chlorine yield common salt. Sodic hydrate (or sodic oxide) and hydrochloric acid also yield common salt. Consider in what respect the formation of common salt differs from that of other salts—for example, magnesium sulphate.

Ex. 51. *Volatile and Fixed Alkali.*—Compare the formation of the ammonia with that of the sodium salts.

Ammonia, gas, water, and sulphuric acid yield ammonic sulphate.

Ammonia, gas, water, and nitric acid yield ammonic nitrate.

Ammonia, gas, water, and hydrochloric acid yield ammonic chloride.

Sodic hydrate and sulphuric acid yield sodic sulphate.

Sodic hydrate and nitric acid yield sodic nitrate.

Sodic hydrate and hydrochloric acid yield sodic chloride.

Compare the two salts of each acid and determine whether volatile or not. Heat solutions of each of the ammonia salts with a solution of caustic soda and test the gas given off with red litmus paper and with great caution by the smell.

11. Copper.

Ex. 52. *Distinctive Characters.*—The metal is best used in a very thin, flexible sheet, easily cut with a pair of scissors. Let the student first study and describe the properties of the metal, determining its specific gravity in the usual way, and comparing its colour, hardness, toughness, malleability, etc., with the similar qualities of zinc. Let him harden on an anvil and anneal with heat. Let him next heat in separate test tubes a few bits of the metal with nitric, hydrochloric, and sulphuric acids, using both strong and weak acid. Compare with zinc. Observe behaviour, to be interpreted beyond.

Ex. 53. (a) *Reduce Oxide with Hydrogen Gas and reoxidize in the Air.*—Introduce into a combustion tube about twenty grammes of black oxide of copper. Take the tare on the balance. Mount on tube furnace, connecting one end with a hydrogen generator and the other end with a small U tube kept cool with ice. Pass now a slow current of hydrogen gas over the powder, and after the gas has expelled the air heat the combustion tube to low redness. Observe that the powder takes on the colour and lustre of metallic copper and that water collects in the tube. After the reduction is complete, dismount the apparatus and reweigh the tube.

(*b*) Remount on the furnace the combustion tube with its contents as left in the last experiment. Leave one end open to the air and connect the other with the gasometer before described (note to Ex. 45), which, having been previously filled with water and the overflow opened into a sink at a lower level, will act as an aspirator. Regulate by pressure tap so that air will be drawn through the tube not faster than about two bubbles a second, and then heat the reduced copper to redness. After collecting one or two litres of gas close and dismount the gasometer, while at the same time connecting the combustion tube with an aspirator pump to hasten the process. Continue heating the tube until the powder has acquired a uniform black color. Then allow to cool, dismount, and reweigh. Meanwhile transfer the gas collected to gas bottles over a pneumatic trough and seek to recognize the substance. Interpret all the phenomena observed. Compare Ex. 24 (*b*) and Ex. 42. Keep the oxide of copper in the tube for another experiment.

(1) Water consists of hydrogen and oxygen. The production of water from hydrogen gas implies what? The formation of water is attended with the change of the black powder to metallic copper. Of what must that powder consist?

(2) Whence comes the nitrogen collected in the gasometer? How may free oxygen, or substances holding oxygen loosely united, be expected to act on free copper? How may

oxide of copper be expected to act on substances containing hydrogen or carbon ?

Ex. 54. (*a*) *Salts of Copper.*—Take three test tubes holding a few cubic centimetres of dilute sulphuric, hydrochloric, and nitric acid respectively. Dissolve in each black oxide of copper, adding in very small quantities so long as solution is obtained on boiling. Evaporate on watch glasses and crystallize the products.

Oxide of copper and sulphuric acid give copper sulphate (salt).

Oxide of copper and nitric acid give copper nitrate (salt).

Oxide of copper and hydrochloric acid give copper chloride (salt).

(*b*) Take a few cubic centimetres of the solution of copper in nitric acid—Ex. 52 or 44 (*a*). Secure the ready solution of copper in sulphuric and hydrochloric acids by adding a few drops of nitric acid in each case; evaporate to dryness (not over 100°), dissolve the residue in the smallest possible amount of water, and allow to crystallize in a warm place. Compare products with those obtained in (*a*). Lastly, heat copper nitrate in a porcelain crucible to low redness, until fumes cease, then to bright redness. When cold pulverize the residue in a mortar and examine the black powder left. Seek to interpret now the phenomena observed when attempting to dissolve the metals.

(1) What is the difference between the action of copper and zinc on dilute sulphuric acid? This difference can be explained by the thermal relations of the metals, but must here be accepted as a fact. Why should you expect that the addition of nitric acid would secure the ready solution of copper in both sulphuric and hydrochloric acids? What double part does nitric acid play in dissolving copper? Compare Ex. 44 (*a*) and also the additional fact that when in this preparation of nitric oxide the temperature is allowed to rise too high and the action to become violent the product is chiefly, or even altogether, nitrogen gas.

12. Iron.

Ex. 55. *Distinguishing Properties.*—The metal is best used in the shape of wrought-iron-wire nails or tacks of various sizes, also in fine powder "iron by hydrogen." Let the student first study and describe the properties of the metal, as in the case of copper (Ex. 52). He should then find its specific gravity, using a method applicable in many cases. Take a small vial or flask (ten to twenty cubic centimetres capacity) and make a mark on the narrow neck. Fill the bottle to the mark with water and tare it on the balance. Select some iron nails as large as will conveniently be held by the bottle. Place about twenty grammes of these nails at the side of the bottle, and take the exact weight. Remove the bottle from the pan and drop in the nails one by one, taking care to avoid entangling bubbles of air. Wipe off the water which runs over, and with a small roll of porous paper reduce the level in the neck to the mark. Replace on the pan and reweigh, when the loss of weight is obviously the weight of water displaced by the nails. In this connection the qualities and relations of the different kinds of iron—wrought iron, cast iron, and steel—should be explained by the teacher, and the prominent features of the metallurgy of iron might appropriately be discussed.

Ex. 56. (*a*) *Burning of Iron.*—If the iron powder is sufficiently fine it will burn on the pan of a balance like magnesium (Ex. 48), and the increase of weight may be found. If too coarse to burn in this way, the iron powder will burn brilliantly by sprinkling it through the flame of a Bunsen burner held obliquely. The residue may be collected on a large sheet of paper held obliquely. A more striking experiment of burning a watch spring in oxygen gas should be made by the teacher. After removing the temper of the steel by heating in a lamp flame, the spring can be coiled into a spiral. Tip one end with sulphur like a match, and hang the other from a wire, which should be slid through a cork, closing the tubulature of the jar as fast as the oxygen is exhausted. Protect the exposed face of the cork with metallic foil, and then light the iron match and plunge it into the gas. The experiment is best made in a tubulated bell jar standing over water into which the oxide of iron melted by the flame falls in drops. These melted globules would crack the bottom of a glass jar unless protected by sand.

(*b*) Weigh out in a porcelain crucible five grammes of the iron powder. Ignite with frequent stirring (use iron wire) until completely oxidized. Reweigh when cold, and interpret the result.

(*c*) Weigh out five grammes of iron powder in a shallow porcelain dish. Keep the powder moist with water until it is wholly converted into rust. Then allow to dry in the air and weigh again. Save a part of the residue for further use.

(1) Why is the weight of the rust in (*c*) greater than that of the red powder formed in (*b*)? Devise an experiment for testing your inference.

Ex. 57. (*a*) *Iron and Sulphuric Acid.*—Iron dissolves in dilute sulphuric acid like zinc or magnesium, with rapid evolution of hydrogen gas. Repeat Ex. 25, using 5 grammes wrought-iron tacks (instead of zinc clippings), also 10 grammes strong sulphuric acid diluted with 30 grammes of water. Use a 100-cubic-centimetre flask. Collect the gas as before, and compare with the previous product. Observe precautions previously given (note to Ex. 27) in regard to the preparation of hydrogen gas.

(*b*) After the evolution of gas has ceased, remove the outlet tube from the flask and provide a tightly fitting cork. Boil down the residual solution to about one half, cork the flask while still full of steam, set aside and allow to cool. Examine the crystals which form. They are ferrous sulphate. Interpret the phenomena observed. What is the source of the hydrogen? Compare Exs. 47 and 48. As evidence bearing on this point, it should here be stated that iron forms with oxygen an oxide containing less oxygen than that obtained in Ex. 56 (*b*), but so difficult to prepare and keep as to be unsuited for class experiments. This oxide dissolves in dilute sulphuric acid without evolution of hydrogen, yielding the same ferrous sulphate. In this connection, and as a step towards the answer to the above question, let the student review the relations of magnesium, magnesium oxide, and magnesium hydrate towards sulphuric acid.

Ex. 58. *Two Classes of Iron Salts.*—It is by no means true of all metallic oxides that they will dissolve directly in acids to form salts without the intervention of any other agent or the formation of any other product. The oxides that do sustain this relation to acids, like most of those we have studied, have been called for distinction sake salifiable oxides. Most of the metals, like sodium, magnesium, and zinc, form but one salifiable oxide; but there are a number of metals which yield two such, and it is a remarkable fact that the two classes of salts thus formed differ widely from each other in their properties and relations. Iron is an example in point, and the two classes of salts thus formed are distinguished as ferrous and ferric salts. The green transparent crystals (green vitriol or ferrous sulphate) formed by dissolving iron in dilute sulphuric acid is a ferrous salt, and the same product, as has been said, also results when the above-mentioned ferrous oxide is dissolved in the same acid. The oxide formed in Ex. 56 (*b*), called ferric oxide, is also salifiable, but when once ignited, as in that experiment, dissolves in acids with difficulty. The hydrate formed by slow oxidation in the air in contact with water dissolves readily. To substantiate that the ferric salts thus formed are essentially different from the ferrous salts, dissolve a portion of the residue from Ex. 56 (*c*) in dilute sulphuric acid, evaporate nearly to dryness on a watch glass, and try to crystallize the residue. Compare this ferric sulphate with the ferrous sulphate before made.

Ex. 59. (*a*) *Iron and Sulphur.*—Thoroughly mix 5 grammes of iron powder with 2·8 grammes of flowers of sulphur. Save a small portion of the mixture for comparison; heat the rest in a small flask (150 cubic centimetres) until the mass glows. After cooling remove with a rod a few grains of the product and compare with the mixture. (Use microscope and magnet.) The black product is called iron sulphide.

(*b*) *Hydrogen Sulphide.*—Leaving residue in the flask, to which has been fitted cork and outlet tube, pour in 10 grammes of strong sulphuric acid mixed with 50 grammes of water. After air has been driven from the flask, collect the first portions of the escaping gas in large test tubes over hot water. Ignite the gas at the open mouth of these test tubes, and observe the colour and odour produced by the flame. Let the rest of the escaping gas bubble up through cold water in a glass-stoppered bottle, and when action is exhausted withdraw outlet tube, stopper the bottle, and set aside for future use. Boil now contents of flask as in Ex. 57 (*b*), and set aside to crystallize. These crystals will at once be recognized as ferrous sulphate, and the nauseous-smelling gas, which is, to a limited extent, soluble in water, is called hydrogen sulphide. It is one of the most important of chemical reagents. As this gas is not only nauseous, but also to some extent poisonous, this experiment must be made under a hood or in the open air. Interpret all the phases.

(1) Iron sulphide obviously consists of iron and sulphur. Iron dissolved in dilute sulphuric acid yields ferrous sulphate and hydrogen gas. Ferrous oxide dissolved in the same acid yields also ferrous sulphate, but no free gas, because the hydrogen, otherwise formed, unites with the oxygen of the

oxide to form water. Ferrous sulphide dissolved in the same acid yields, again, ferrous sulphate and a nauseous-smelling gas. What must be the composition of this gas? Do the phenomena observed when the gas burns confirm your inference?

CHAPTER II.

GENERAL PRINCIPLES.

13. Province of Chemistry.

At this stage of his study the student, having become familiar with the distinctions implied by the word "substance," and having acquired some knowledge of chemical phenomena, is prepared to understand what is the province of the science of chemistry. *Chemistry comprises and classifies our knowledge of those phenomena which imply a change of substance.* The science of physics, on the other hand, deals with phenomena which do not necessarily imply a change of substance; and hence the distinction between chemical and physical changes. This distinction should be illustrated by the teacher from the experiments already made by the student.

In every chemical change one or more substances, called the *factors*, change into one or more other substances, called the *products;* and it is a primary object in the study of chemistry to learn what are the factors and what are the products of every process that comes under notice.

Substances may be mixed with one another, like the ingredients of gunpowder, or one substance may be dissolved in another, like salt in water, without undergoing chemical change; but in all such cases the qualities of the original substances may be recognized in the mixture or solution. Hence the very broad distinction between a mixture, or a solution, and a chemical combination, which the following experiments will illustrate.

Ex. 60. *Mixture and Chemical Compound.*— Mix together in a mortar as intimately as possible 3·26 grammes of zinc dust with 1·60 gramme of flowers of sulphur. First examine a small amount of this powder under a microscope of sufficient power to show the yellow grains of sulphur lying side by side with the metallic grains of zinc. Make with the rest of the mixture a conical pile on a square of asbestos paper, and apply the flame of a match. A chemical change ensues, marked by a brilliant deflagration; and as the result there is left on the paper a white powder, which is a compound of zinc with sulphur, and is called sulphide of zinc. Here there were two factors of the chemical change, zinc and sulphur, and one product, sulphide of zinc. Examine with the microscope this product, and no traces can be seen either of zinc or of sulphur. The deflagration was a manifestation of the heat evolved by the chemical change; and in every chemical change there is either a setting free or an absorption of energy, usually as heat. But this last feature of a chemical process will be considered later by itself. Compare Ex. 59 (*a*).

Ex. 61. *Physical and Chemical Solution.* — Take two portions of one gramme each of sodic carbonate. Dissolve one portion in three cubic centimetres of water, evaporate to dryness slowly, and compare the residue with the original salt in appearance, crystalline form, and taste. Dissolve the second portion in dilute hydrochloric acid, evaporate, and compare. What are the factors and what are the products of the chemical change in the second case? Notice that water is the medium of the chemical process, and dissolves one of the products; so that we have here both a chemical change and a simple solution. The same is true when zinc dissolves in dilute sulphuric acid (Ex. 47), or when copper dissolves in dilute nitric acid — Ex. 44 (*a*) — and the double use of the term "solution" must be made clear. What are the factors and what are the products in the two cases of chemical solution last cited? In the same way the teacher should review the experiments which the student has made, and point out what are the factors and what are the products in each case.

14. Fundamental Laws.

In every well-marked chemical change three fundamental laws are observed, and these are called the law of conservation of mass, the law of definite proportions by weight, and the law of definite proportions by volume.

Ex. 62. *Law of Conservation of Mass.*—Repeat Ex. 24, but after adjusting the apparatus with the bit of phosphorus in the spoon and fastening the cover, balance the jar on the pan of the balance with a second jar of the same volume and such additional tare as may be needed. Ignite now the phosphorus with a burning-glass, and after the chemical action, when the jar is cold, replace it on the balance. If the jar was tight there will have been no change of weight. Hence it must be that—

The sum of the weights of the products of a chemical change is exactly equal to the sum of the weights of the factors.

We may conceive of any chemical process as taking place in an hermetically sealed space—indeed, the earth is essentially such a space—and hence this law must be universally true. The result of this experiment might be anticipated, and it may therefore be thought unnecessary; but its very form will make evident to the student that the law of conservation of mass is in harmony with general principles which he already recognizes.

Ex. 63. *Law of Definite Proportions by Weight.*—Take five grammes of sal-soda (crystallized sodic carbonate), selecting material that has not effloresced; dissolve in dilute hydrochloric acid, as directed in Ex. 61, taking care to avoid loss during the effervescence; evaporate to dryness, and weigh the residual salt; calculate the ratio of the sal-soda used to the salt produced; repeat the same determination with ten grammes of sal-soda, and within the limits of experimental error the ratio will be the same as obtained before, and so would it be whatever the amount of sal-soda employed. In this chemical change the factors are sal-soda and hydrochloric acid, while the products are common salt, carbonic dioxide, and water. The last two, being volatile, escape during the effervescence and subsequent evaporation. By this experiment we have proved that the proportion between the weight of the sal-soda and the weight of the common salt is definite, and it could readily be shown experimentally that the proportion between any two of the five substances involved in this chemical change was equally definite. So of any other well-marked chemical change, and hence the general law that—

In any well-marked chemical change the rela-

tive weights of the several factors and products are definite and invariable.

Here, again, the result might have been anticipated, for it only amounts to finding that if we use twice as much sal-soda we shall obtain twice as much common salt, which might seem self-evident; and this consideration will show that the law of definite proportions by weight is in entire harmony with principles universally recognized.

Ex. 64. *Law of Definite Proportions by Volume.*—This law, sometimes called the law of Gay-Lussac, may be thus stated:

In any well-marked chemical change the relative volumes of the aëriform factors or products, if measured under the same conditions, bear to each other a simple numerical ratio.

It has already been illustrated by several experiments, which it is unnecessary to repeat. Thus it was shown by Ex. 32 that when oxygen combines with sulphur to form sulphurous oxide the volume of this sole product is the same as the volume of the oxygen gas used. A similar relation appeared when oxygen united with carbon to form carbonic dioxide in Ex. 24 (*c*). Again, when, in Ex. 40 (*a*), carbonic dioxide united with more carbon to form carbonic oxide the volume of the gas was doubled. A still more striking illustration of the law is to be found in the fact that two volumes of hydrogen gas combine with one volume of oxygen gas to form two volumes of vapour of water, all measured, of course, under the same pressure and at a temperature above the boiling point of water. The experiment is easily made with a form of eudiometer invented for the purpose by Hofmann and sold by all the dealers in chemical apparatus, and it should be shown to the class if possible.

15. Compounds and Elements.

The student can not have performed the experiments heretofore described without himself drawing the inference in certain cases that the products have been formed by the union of two or more factors, and in other cases that the products have resulted from the breaking up of a factor into simpler parts. Hence come the fundamental conceptions of composition and decomposition, of synthesis and of analysis, as we have previously called them. Our judgment in any case depends not only on the circumstances of the experiment, but also on a comparison of the weights of the products with those of the factors from which they were formed. Thus, in Ex. 24 (*a*), it is perfectly evident from the conditions of the experiment that the white product results from the union of phosphorus and oxygen. If now in addition we could weigh the white product and find that its weight was exactly equal to that of the phosphorus and oxygen used, the proof of its composition would be complete. So also when, in Ex. 16, we pass a current of electricity through water and see oxygen and hydrogen gases escaping from the platinum poles of the apparatus, and notice that everything else remains unchanged, we conclude that the two gases must come from the water and are the products of its decomposition; but we do not have absolute proof until, as in Ex. 53 (*a*), we pass hydrogen over oxide of copper and find that the weight of the water formed is exactly equal to that of the hydrogen and oxygen which have disappeared. In like manner, our knowledge of the composition of other substances is the result of our knowledge of chemical processes, which has been accumulated during long years of study. As the total result of this study, we may say that while the larger number of substances which we handle may be decomposed or analyzed, there are about seventy known substances which can not be broken up into simpler parts, and these we call elementary substances. An elementary substance differs from other substances only in this, that it enters into all chemical changes as a whole, and we know of no chemical process in which it becomes divided. It does, however, enter into union with other substances; and, speaking in general, we may by the combination of the elementary substances reproduce all the forms of matter with which we are acquainted. The systematized knowledge of the methods, whether analytical or synthetical, by which the composition of bodies has been determined is a very important branch of chemical science, known under the name of chemical analysis; and the subject is subdivided into qualitative and quantitative analysis, according as the object in view is to determine solely the nature or the proportion of the ingredients. In either case the analysis may be either ultimate or approximate. It is ultimate when we seek for the elementary substances of which the compound consists. It is approximate when we look for the simpler products (for the most part acids and metallic oxides) into which the complex material may be primarily divided. The following experiments will give a general, but necessarily a very imperfect idea of the manner in which the results are reached:

(1) A list of the elementary substances will be found in the table at the end of this book, and this should be carefully examined by the student in reference to the substances he has met with in the course of his experiments. Which of these are elements? The student should make a list of the elementary substances with which he has become familiar. Can an elementary substance be told by its external characters? Is there not a class of bodies which are uniformly elementary?

16. Qualitative Analysis.

Ex. 65. *Analysis of a Silver Coin.*—Dissolve a ten-cent coin in 5 cubic centimetres of pure, strong nitric acid, diluted with its own volume of water. Dilute to 50 cubic centimetres. Add hydrochloric acid to hot solution so long as a precipitate is produced. Filter, wash thoroughly (three times) with water, dry precipitate. Transfer to glass combustion tube connected with hydrogen generator as in Ex. 53 (*a*). Interpose chloride-of-calcium tube between generator and combustion tube. Allow hydrogen to flow through the apparatus long enough to expel the air; then heat the combustion tube, and continue until reduction is complete. Test gas evolved from outlet, which should be bent downwards. Preserve silver. Add a few drops of sulphuric acid to blue filtrate and evaporate (under hood) to get rid of the volatile acids; dilute to 10 cubic centimetres and insert a strip of zinc.

(1) What is the gas evolved during the reduction? What is its composition? What must be the composition of the precipitate? Does the formation of this precipitate conform to any general principle—Ex. 14 (3). On what circumstances does the separation of copper form silver in this experiment? How can you be sure that the copper obtained came from the coin and not from any of the accessory materials employed?

Ex. 66. (*a*) *Analysis of Marble.*—Heat two grammes of marble dust in a small iron crucible over a blast lamp, so long as the material continues to lose in weight. The residue easily recognized as lime must be one of the constituents. Add water, test with litmus paper, and compare with a mixture of marble dust and water. What inference would you draw from Ex. 48 in regard to the probable constitution of such white powders? We know that magnesium has a very strong attraction for oxygen (Ex. 48), and therefore, to test this inference heat over a lamp in a small ignition tube two decigrammes of lime in powder mixed with one decigramme of magnesium powder. When cold add water, and shake up with residue. Note the evolution of hydrogen gas and the production of lime water. The metal liberated, which decomposes water, is calcium. Hence lime consists of calcium and oxygen.

What was driven off from the marble dust by heat? To show this, let the teacher procure half a metre of small iron gas pipe. Close one end by welding. Drop in ten grammes of marble dust and shake down to the closed end. Mount in a Fletcher gas furnace and connect the open end with a pneumatic trough. Heat to a full white heat, avoid excess of illuminating gas at the burner lest it diffuse through the tube. Collect and examine gas evolved. It will not support combustion, it dissolves in water, and feebly reddens litmus paper. It is obviously carbonic dioxide. Of what does carbonic dioxide consist? Take a length of magnesium ribbon. Ignite and plunge the burning end in a jar of carbonic dioxide. Observe the separation of carbon. What is the necessary inference in regard to the composition of carbonic dioxide? Discuss all points of this evidence.

(*b*) *Synthetical Confirmation.*—Review in this connection Ex. 39 (*b*), and repeat on a small scale with the lime water obtained by the action of calcium on water. Discuss the phenomena as synthetical evidence of the composition of marble. Explain the action of hydrochloric acid on marble, bringing in contrast the two facts—

Marble and hydrochloric acid yields calcic chloride and carbonic dioxide.

Lime and hydrochloric acid yield calcic chloride.

Confirm these facts experimentally and draw your own inferences. As will hereafter appear, the facts as above stated are not complete statements, since in both processes water is also formed, which in qualitative experiments escapes notice by mixing with the mass of liquid, acting as the medium of the chemical change. Still in the present case the fact overlooked was not material, and, since in experimental science we can frequently draw correct conclusions from similar incomplete evidence, this experience may teach a valuable lesson. It was from exactly this evidence that the proximate composition of marble was first inferred by Dr. Black, of Edinburgh, a century ago.

Synthetical processes are often of great value in confirming analytical results, and give frequently the most direct and efficient means of finding out the composition of a material. As commonly used the term chemical analysis includes all methods of establishing the chemical constitution of substances, whether synthetical or strictly analytical.

Ex. 67. *Chemical Tests.*—In the examples of analysis given above, we have actually separated the elementary substances of which two familiar bodies consist. But such a separation is not always practicable or necessary, and we can generally discover the constituents of a substance by applying certain characteristic tests.

Take four short lengths, not over 50 millimetres long, of platinum wire and make a loop at the end of each; melt into the first of these loops chloride of sodium; into the second, chloride of potassium; into the third, chloride of strontium; and into the fourth, chloride of barium. Hold the loops successively in the flame of a Bunsen lamp, and notice the colours which they impart to it; and if a spectroscope is accessible, examine the coloured flames with this instrument. Almost any preparation of sodium, potassium, strontium, or barium would produce the same effect; and these characteristic colors, or still better the corresponding bands seen in the spectroscope, are indications, or, as we usually say, tests of these metals. For another illustration, take five test tubes; in the first dissolve a small amount of zinc dust in acetic acid; in the second, a bit of white arsenic in hydrochloric acid; in the third, a bit of antimony in hydrochloric acid to which has been added a drop of nitric acid; in the fourth, a bit of iron in hydrochloric acid; in the last, a bit of lead in weak nitric acid. In each case use a bit of metal no larger than a pin's head, and dissolve in the least amount of acid possible; half fill the test tubes with water; add to the first three the solution of sulphide of hydrogen obtained in Ex. 59 (*b*); to the fourth, a few drops of a solution of ferrocyanide of potassium; and to the last, a few drops of a solution of potassic chromate. The characteristically coloured precipitates obtained under the conditions present are in each case tests of the several metals. So, in general, it is not necessary to isolate an acid, a metallic oxide, or an elementary substance, in order to prove that it is present or absent in a given case, but only to use the proper tests in the right way; and the works on qualitative analysis teach us in what order and under what conditions the proper tests should be applied. The practice of qualitative analysis affords a most admirable training in the methods of inductive reasoning.

17. Quantitative Analysis.

Ex. 68. *Analysis of Potassium Bromide.*— The analysis made in Ex. 65 may be made quantitative by first weighing the coin and afterwards weighing the silver and copper obtained. Of course if the coin consists of nothing else, the sum of the weights of the two metals ought to exactly equal the weight of the coin, and such a coincidence would go far to establish the accuracy of our work. In the analysis as above made only a very rough approximation to equality could be expected, but by more accurate methods such a confirmation of the work could be almost absolutely secured. In general, however, in order to determine the relative proportions of the different constituents in a compound, it is rarely practicable to separate the ingredients and weigh the several amounts. The method is to transfer each ingredient to some new combination which can be formed without loss, weighed with accuracy, and the composition of which through previous analyses is absolutely known. Take the simplest case. We wish to analyze common salt, which is known to consist wholly of chlorine and sodium. Neither of these are ingredients which can be accurately separated and weighed. So, with a carefully weighed quantity of salt, we prepare chloride of silver by a process in which we are sure that every particle of chlorine has been transferred from its previous combination with sodium to a new combination with silver. Chloride of silver is a substance which can be collected and weighed with perfect precision. It has previously been accurately analyzed over and over again, and by referring to tables we find what fraction of the weight of chloride of silver thus found consists of chlorine, and then a very simple calculation gives the weight of chlorine sought. If the salt is absolutely pure the rest of the weight taken consists of sodium, and the amount of sodium could not be determined so accurately in any other way. Obviously, the analysis of common salt rests back on the analyses of chloride of silver previously made and recorded; and so in most cases our analyses of to-day rests back on the work of those who have gone before us. After these relations have been explained, let the student weigh in a small beaker glass exactly one gramme of pure potassic bromide,* and dissolve the salt in about

---

* With the rude manipulation here expected, potassium bromide will give more precise results than common salt and illustrates equally well the general methods of quantitative analysis. Potas-

twenty-five cubic centimetres of water. Weigh in a similar beaker one and a half gramme of silver nitrate and dissolve in an equal amount of water. Pour now with constant stirring the first solution into the second, rinse the beaker, wash in the last drops, and allow to stand until the precipitate fully settles. Collect on a tared filter, wash dry, and weigh. It is known from previous work, and can be found by reference to any work on quantitative analysis, that every gramme of silver bromide contains 0·4255 gramme of bromine; and practically an analyst would always assume that this value was given and at once calculate the amount of bromine in the weight of the silver bromide he had obtained, and this would be the amount in the weight of the bromide of potassium he had taken for analysis. To show the student, however, that results thus obtained rest back on previous analyses, let him make this additional determination, not that he can compete with the old work, which has been often repeated with the greatest care in order to establish fundamental data for just such uses as are here indicated, but to the end that he may realize the actual relations in most problems of quanitative analysis. Weigh out exactly one gramme of pure metallic silver,

---

sium bromide is a familiar medicinal salt, consisting of potassium and bromine, two elementary substances closely allied respectively to sodium and chlorine.

place in a small beaker, and dissolve in about two cubic centimetres of strong nitric acid diluted with four cubic centimetres of water, and add twenty-five cubic centimetres of water. In a second beaker dissolve in twenty-five cubic centimetres of water 1·2 gramme of potassic bromide, and then proceed as before. From the result calculate the weight of bromine in one gramme of silver bromide, which should be, within the limits of error, the same as the value given above. The student should now calculate from his own results the per cent of bromine and of potassium in potassic bromide, and should have practice in similar calculations until he is familiar with the usual manner of stating the results of analysis in per cent.

(1) It is known that pure common salt consists wholly of sodium and chlorine, and also that one gramme of silver chloride contains 0·2474 gramme of chlorine. In one determination 0·5723 gramme of salt gave 1·4038 gramme of silver chloride. Calculate the percentage composition of common salt.

Ans. Chlorine, 60·69
Sodium, 39·31
―――――
100·00

(2) It is known that pure crystallized cane sugar consists wholly of carbon, hydrogen, and oxygen. By the usual process of organic analysis, 0·2569 gramme of sugar gave 0·3966 gramme of carbonic dioxide and 0·1487 gramme of water. It is known that one gramme of carbonic dioxide contains 0·2727 gramme of carbon, and one gramme of water 0·1111 gramme of hydrogen. What is the percentage composition of cane sugar ?

Ans. Carbon, 42·11
Hydrogen, 6·43
Oxygen, 51·46
―――――
100·00

(3) Obviously the above processes assume a qualitative knowledge of the composition of the substance analyzed, and so, in general, quantitative analysis implies a previous qualitative analysis. Indeed, the process of determining the amount of an ingredient present must constantly be varied according as it is associated with different substances, and a large knowledge is required in order to meet the conditions in any case and secure accurate results. Thus quantitative analysis becomes a distinct and widely extended branch of chemical study, and it is the chief work of the practical chemist.

CHAPTER III.

MOLECULES AND ATOMS.

18. Molecular Theory.

THE theory of the new science of thermo-dynamics assumes that the material of aëriform bodies is not continuously distributed through the spaces they seem to fill, but consists of a vast number of exceedingly minute particles in rapid motion to and fro, constantly rebounding from one another, or from the walls of the containing vessel. These minute particles are called molecules, and the phenomena of heat are supposed to be manifestations of their moving power. The molecules of the same substance, of water, for example, are supposed to have the same weight—in fact, to be alike in every respect; while the molecules of different substances are as unlike as the substances themselves. This theory has been worked out mathematically with great ability, and the phenomena of nature have been found, in a most remarkable manner, to conform to the deductions of mathematical analysis. Of these deductions one of the most remarkable is, that—

Equal volumes of all gases or vapours, measured under the same conditions, contain the same number of molecules.

This deduction is usually called the law of Avogadro; and if we accept the fundamental conception of molecular structure we must also accept this inference which it involves. The modern theory of chemistry accepts the law of Avogadro as a fundamental principle, and builds upon it a large superstructure.

The law of Avogadro does not absolutely hold except when the material is in a perfectly aëriform condition. It is only approximately true in the case of dense gases or vapours under the pressure of the air, and near the point of condensation the deviations are sometimes very marked. It has no reference whatever to liquids or solids. These forms of matter are supposed also to consist of moving particles; but, if so, the molecules must be variously compacted, and their motions otherwise circumscribed than in the aëriform state.

19. Physical Method of Determining Molecular Weights.

If the law of Avogadro is true, the molecular weight of a substance must be proportioned to its specific gravity in the state of gas or vapour. If we take hydrogen gas as the unit of reference for the specific gravity, and the molecule of hydrogen gas as the unit of reference for molecular weights, then the number which expresses the specific gravity of a substance in the state of gas or vapour would also express the molecular weight of that substance. For considerations which will shortly appear, half the weight of the molecule of hydrogen has been taken as the unit of molecular weight, so that the molecule of hydrogen gas weighs two of the assumed units; and hence, on this system, the molecular weight of any substance is found by doubling its specific gravity taken in the aëriform state, and referred to hydrogen gas as the standard.

The physical method of determining molecular weights, therefore, reduces itself to finding the specific gravity of a substance when in the condition of gas or vapour with reference to hydrogen. The substance must be in the condition of gas or vapour, and the method is only applicable to those bodies which are naturally aëriform, or which can be volatilized at temperatures within control without undergoing decomposition.

Ex. 69. *Density of Hydrogen.*—Use a flask not exceeding 100 cubic centimetres capacity. Cork tightly, and connect through cork a small chloride of calcium tube, so proportioning the parts that the flask will stand on the pan of the balance. Add to the flask 20 cubic centimetres of strong hydrochloric acid and an equal volume of water. Weigh out closely 5 grammes of sheet zinc that has been carefully cleaned. Place this at the side of the flask on the scale pan and take the tare as closely as possible. Removing now the flask from the balance, connect, by means of a flexible-rubber connector, the chloride-of-calcium tube with the inlet tube of the gasometer before described (note to Ex. 45), which should be large enough to hold two full litres of gas. When all is ready withdraw the cork, drop in the zinc, and quickly recork the flask. Wait until the evolution of hydrogen has altogether ceased, then shut off the gasometer and disconnect the flask. Before replacing it on the balance pan withdraw the cork for a few minutes to give the hydrogen gas in the interior time to diffuse into the air. The loss of weight is obviously the weight of the hydrogen set free. To find the·volume of this hydrogen transfer the gas from the gasometer in successive portions to a litre measure over a pneumatic trough. In reading the volumes take care that the level of the water is the same inside and outside the glass and avoid warming with the hands. This volume should be corrected for the tension of aqueous vapour (Ex. 22) and reduced to standard conditions, observing, for the purpose, the height of the barometer and the temperature of the water in the trough. We have, then, the weight of a measured volume of hydrogen, and can easily calculate the weight of one litre. With such appliances as are here assumed the process is accurate within about 5 per cent. On account of its great lightness the exact determination of the density of hydrogen gas is a difficult problem, and for the purpose of calculating the specific gravity of other aëriform bodies referred to hydrogen as unity, we will assume the value generally received, 0·0896, and for the specific gravity referred to air, 0·0692. (Compare Ex. 17.)

Ex. 70. *Specific Gravity of Carbonic Dioxide.*—Take a quart tin can, with narrow neck fitted with selfsealing stopper, and, having measured its exact contents, as in Ex. 17, carefully clean and dry it. Place it, open, on the balance-pan, and equipoise it with a second can of the same size and pattern, tightly sealed. Observe the thermometer and barometer at the time the equilibrium is established. It will be obvious now that if we calculate the weight of air which the open can holds at this temperature and pressure (Ex. 17) and add an equal weight to the pan carrying the open can, we should have what would be the exact tare if the can were sealed with all the air exhausted from the interior. Moreover, since the can and its counterpoise displace the same volume of air, it is also obvious that this equilibrium would not be disturbed by any changes in the atmosphere. If, therefore, we fill the can with any gas—for example, carbonic dioxide—the increased weight will be simply the weight of this gas. Remove then the open can, fill it with carbonic-dioxide gas by displacement, and seal it, observing the thermometer and barometer at the moment the can is closed. Determine the increased weight; and this is the weight at the last observed temperature and pressure of a volume of carbonic-dioxide gas equal to the known capacity of the can. From this value calculate what would be the weight of the same volume if the thermometer marked 0° and the barometer stood at 30 inches. The density of carbonic-dioxide gas — that is, the weight of a litre under the standard conditions of temperature and pressure—is then found by dividing the weight of the gas by the capacity of the can. The density of carbonic dioxide divided by the density of air gives the specific gravity of carbon dioxide referred to air; or, if divided by the density of hydrogen, the specific gravity referred to hydrogen; and these values are the same for all temperatures and pressures. Why?

(1) What is the molecular weight of carbonic dioxide?

(2) The specific gravity of nitrous-oxide gas referred to hydrogen is 22·04. What is its molecular weight?

(3) The specific gravity of cyanogen gas referred to hydrogen is 26·06. What is its molecular weight?

(4) The specific gravity of oxygen gas referred to hydrogen is 16. What is its molecular weight?

Ex. 71. *Specific Gravity of Vapours.*—Find the molecular weight of alcohol, ether, chloroform, or ethylene bromide, by determining in either case the specific gravity of the vapour of the substance by the following method:

Determine the volume of one of the bulbs provided for the purpose,* and, having thoroughly dried both the interior and the exterior surface, seal the shorter tubulature. Select a second bulb of approximately the same size, and, having sealed both of its tubulatures, use it to equipoise the first, completing the tare as convenient. Observe now the thermometer and barometer, and calculate the weight of the dry air which fills the open bulb—Ex. 17 (1). Add this weight to the pan holding the first bulb; and, as thus loaded, the balance would be in equilibrium were the glass vessel completely exhausted. Moreover, this constructive equilibrium will not be disturbed by any atmospheric changes (Ex. 70). Introduce now into the first bulb about 50 cubic centimetres of the volatile liquid under examination. Hang the bulb above the water in an ordinary tea-ket-

---

* These bulbs hold about 400 cubic centimetres, and are blown in a mould so as to secure a uniform size. They have a long, narrow stem and opposite to the stem a short and still narrower tubulature.

tle sufficiently capacious for the purpose, with the longer tubulature projecting through a cork fitting a hole in the cover. Boil the water under the bulb so that the steam surrounding the glass and escaping by the nozzle of the kettle shall maintain a uniform temperature near 100°. Observe this temperature with a thermometer passing through a cork fitted to a second hole in the cover, and as soon as the current of vapour from the bulb stops seal the tubulature by melting with a blowpipe the glass at the tip. At the same time note the height of the barometer. When cold, replace the bulb on the balance and determine the increased weight above the equilibrium just described. This value is the weight of the vapour which filled the bulb when in the kettle at the temperature and pressure observed. Find next what must have been the slightly increased volume of the bulb when in the kettle, using the formula—

$$V' = V (1 + 0{\cdot}000024 \times \overset{.}{t}{}^{\circ}),$$

and calculate what would be the weight of the same volume of dry air at the temperature and pressure in the kettle. Then the weight of the vapour divided by the weight of the air gives the specific gravity of the vapour referred to air, and this result multiplied by 14·43 gives the specific gravity of the same vapour referred to hydrogen gas. To make sure that the bulb when sealed was full of vapour, break off the tip of the tubulature under water (recently boiled to drive out the dissolved air), when the liquid should rush in and completely fill the interior. If any considerable volume of residual air then appears (more than five or ten cubic centimetres) the determination should be repeated, using more liquid and taking more care to seal the bulb at the right time. If, as is usually the case, the material used is combustible, the right moment is easily caught by lighting the jet of vapour as it issues from the tubulature (after the first violent rush has ceased), and watching the flame as it diminishes. The moment the flame disappears the tubulature should be sealed. In repeating the experiment it is of course unnecessary to alter the tare or disturb the equilibrium if only the tips broken off are kept and returned to the balance pan with the bulb.

(1) The student should now answer the following questions: (1.) What was the *density* of the vapour which filled the bulb when in the kettle at the temperature and pressure noted? (2.) According to what laws does the density of a dry vapour vary when it freely partakes of the temperature and pressure of the surrounding medium? (3.) Does a vapour confined over the liquid from which it rises conform to the same laws? (4.) Why does not the specific gravity of a vapour vary with the temperature and pressure? (5.) Why is the molecular weight of a substance equal to twice its specific gravity in the aëriform state referred to hydrogen gas?

(2) The student should carefully review in this connection **Subdivision 2**, on Air.

(3) The above method obviously only applies to such liquids as boil below the boiling point of water. With less volatile liquids the bulb may be sunk, by means of appropriate apparatus, in a bath of melted paraffine maintained at a constant temperature ; and the specific gravity of the vapours of comparatively fixed bodies has been formed by using globes of porcelain heated in a bath of boiling zinc by means of very powerful furnaces. Quite a different method of experimenting * is better adapted to such cases, but it is beyond the scope of this book.

(4) Sharp results conforming to theory can not be expected by the method here described unless the materials used are of a very high degree of purity. The alcohol must be absolute and the ether or chlorform free from all admixtures. Obviously a process of fractional distillation takes place in the bulb and the impurities may be concentrated in the residual vapour which is weighed. This objectionable feature of the process is avoided in still a third method, applicable only to comparatively volatile liquids in which the volume of vapour formed by a weighed amount of substance is accurately measured under observed conditions. For a description of this process see *loc. cit.* in (3).

20. Chemical Method of determining Molecular Weights.

From the chemist's point of view, the molecule of a substance is the ultimate particle which possesses the qualities of the substance. Molecules may be divided, but if divided the properties of the substance are lost, new molecules are formed, and new substances with new properties appear. In every chemical process the change must take

---

* See author's Chemical Philosophy, pp. 32–37.

place between molecules; that is, one or more molecules of one substance must act upon or must yield one or more molecules of another substance. Hence it must be that in any chemical change the weights of the substances involved must be in the proportion of their molecular weights, or in some multiple of this proportion. In other words, assuming the fundamental conception of molecular structure, the law of definite proportion is a necessary deduction of the molecular theory. And not only is the law of definite proportions a fixed principle of nature, but, moreover, the definite proportions shown by chemical analysis are found to bear a very simple numerical relation to the molecular weights measured by the vapour densities. Thus the facts of chemistry furnish a very remarkable confirmation of the molecular theory of physics. Moreover, the methods of quantitative chemical analysis give us another method of determining molecular weights; for if in any chemical process we can find the quantitative relations between any two of the substances concerned, whether as factors or products, the ratio of these weights must be the proportion of the molecular weights, or else some simple multiple of that proportion; and in most cases we are able to infer from the chemical relations what the multiple is.

Ex. 72. *Molecular Weights of Potassic Chlo-*

rate and of Potassic Chloride.—Weigh in a porcelain crucible (or better, a small platinum crucible) about two grammes of dry potassic chlorate (powdered). Heat to fusion, gradually increasing the temperature as the oxygen gas escapes, taking care to avoid sputtering, and finally heating to low redness for several minutes. Weigh the residue. Make the proportion, As the weight of the oxygen driven off is to the weight of potassic chlorate taken, so is 32, the molecular weight of oxygen gas assumed to be known, to the corresponding weight of potassic chlorate. The weight thus found is known to be two thirds of the molecular weight. What is the molecular weight? Make also the proportion, As the weight of oxygen expelled is to the weight of potassic chloride left, so is 32, as before, to a number which is known to be two thirds of the molecular weight of potassic chloride. What is the molecular weight of potassic chloride?

Ex. 73. *Molecular Weight of Oxalic Acid.*—To a small and light glass flask fit a cork with two perforations; to one of these adapt a small chloride-of-calcium tube, and to the other a short outlet tube. Weigh in the flask about one gramme of crystallized oxalic acid, determining the weight with precision. Dissolve in fifty cubic centimetres of water and add ten cubic centimetres of strong sulphuric acid; allow to cool, and then add one gramme of powdered pyrolusite. After the apparatus is thus mounted take the tare. Then, closing the outlet tube with a small bit of wax, gently heat the flask so long as the evolution of carbonic dioxide continues. Again allow to cool, and when cold remove the wax stopper and suck through the chloride-of-calcium tube so long as the taste of carbonic acid is perceptible. Lastly, determine the loss of weight, which is the weight of the carbonic dioxide formed in this somewhat complex chemical process. Still the principle holds that as the weight of the carbonic dioxide formed is to the weight of the oxalic acid used so is 44, the known molecular weight of carbonic dioxide, to a number which must be a simple multiple or submultiple of the molecular weight of crystallized oxalic acid. In this case the result will be one half of the required molecular weight. Let the student see that in this method it is only necessary to know the relative weights of two of the substances concerned in the chemical process, which may be very complex and in regard to which nothing else need be known. Let him also notice that the chemical method has the advantage over the physical method in that it is applicable to substances which are not volatile. It adopts *the same unit* as the physical method, and refers the unknown molecular weight to a molecular weight previously determined and in the first instance controlled by the physical method; but in this way, step by step, it covers the whole ground. It is far more accurate than the physical method, and practically the physical method is only used to control the results of chemical analysis — that is, to show whether the definite proportions observed are the relations between single molecules or between multiple molecules. Since, however, the chemical method involves the question of multiple ratios, it has its necessary limitations. When the substance under examination is volatile, the results of analysis can, as just said, be controlled by a determination of vapour density. In other cases a study of the chemical process itself gives us the additional information required, but in a way that can not be made intelligible in this connection. Not unfrequently, however, all these means fail, and then the chemist is forced to select between the possible multiples the value which he thinks most probable.

21. Conception of Atoms.

The ultimate particles of substances, called molecules, although far beyond the range of visibility, are by no means inconceivably small, and still less indivisible; and the teacher should attempt to aid the student's imagination by giving the estimates of physicists in regard to the absolute size of these bodies, and showing that the relations between their dimensions and our ordinary standards of magnitude are no more extreme than those we meet with in astronomy, in electricity, and in other branches of physical science. So far are the molecules from being indivisible that it is perfectly evident that they must be divided in almost every chemical change. For example, as we have seen, two volumes of hydrogen gas combined with one volume of oxygen gas to form two volumes of aqueous vapour. Here it is evident that if equal gas volumes contain the same number of molecules, it must be that every two molecules of hydrogen gas combine with one molecule of oxygen gas to form two molecules of water—that is to say, the molecule of oxygen is divided between two molecules of water; or, again, every molecule of water contains one half as much oxygen as the molecule of oxygen gas. We reach the same result in the analysis of water. If we calculate from the results of analysis the percentage composition of water, we find that in 100 parts water contains of hydrogen 11·111 per cent, and of oxygen 88·888 per cent. Further, the specific gravity of aqueous vapour referred to hydrogen is 9, and hence the molecular weight of water is 18. Of this weight, 88·888 per cent, or 16 parts, consist of oxygen. Again, the specific gravity of oxygen gas referred to hydrogen is 16, and therefore the molecular weight of oxygen gas is 32. We know then that—

> One molecule of oxygen gas weighs 32 microcriths.
> One molecule of water weighs 18 microcriths.
> One molecule of water contains 16 microcriths of oxygen.

Hence, as before, every molecule of water contains one half as much oxygen as the molecule of oxygen gas. Obviously we can repeat this calculation with every compound of oxygen in regard to which we know the molecular weight and the per cent of oxygen which the compound contains. If now we arrange the results in a table, as below—

ATOMIC WEIGHT OF OXYGEN.

Compounds of Oxygen.	Observed Sp. Gr.	Weight of Molecule.	Weight of Oxygen in Molecule.
		M. C.	M. C.
Water................	9·00	18	16
Carbonic oxide.	13·95	28	16
Nitric oxide..........	14·97	30	16
Alcohol	23·28	46	16
Ether...	37·32	74	16
Carbonic dioxide......	22·06	44	32
Nitric dioxide.........	24·82	46	32
Sulphurous dioxide....	32·24	64	32
Acetic acid...........	30·07	60	48
Sulphuric trioxide.....	39·87	80	48
Osmic tetroxide.......	128·30	263·2	64
Oxygen gas...........	15·96	32	32

it will appear that in every case the molecule of an oxygen compound contains either sixteen microcriths of oxygen or a simple multiple of sixteen microcriths. The smallest amount of oxygen in any molecule is sixteen microcriths, and this is the weight of what we call an atom of oxygen. The word "atom" is derived from a Greek word meaning indivisible; and this smallest known mass of oxygen, weighing sixteen microcriths, has never been divided. The molecule of oxygen gas consists of two atoms, and of course can be broken in two. We can reason in regard to the compounds of every other elementary substance in precisely the same way and make a similar table,* and in each case we shall find that the weights of a given element in the molecules of its several compounds are simple multiples of a smallest weight, which we take as the weight of the atom of that element. In the case of the compounds of hydrogen the smallest weight is one microcrith, which is the weight of the atom of hydrogen—the smallest mass of matter recognized by science. It is not unreasonable, therefore, that it should be selected as the unit of molecular and atomic weights; and we call this unit by a definite name, a microcrith, in order that the student may associate with the name a real if not a tangible mass of matter. The molecule of hydrogen gas, like the molecule of oxygen gas, contains two micro-

---

* See author's Chemical Philosophy, pp. 42–45, or New Chemistry, pp. 141.

criths, and hence before we attained the conception of the hydrogen atom we described correctly the unit of molecular weight as the half-hydrogen molecule. In this way our conception of atoms and our general knowledge of atomic weights have been reached, and in every work on chemistry the values of the atomic weights adopted are given in tables opposite to the names of the elements. (See table at end of this book.) The student must seek to make clear to his mind the distinction between the conception of the atom and the conception of the molecule. The ultimate particles which retain the qualities of a substance are molecules, and there are as many kinds of molecules as there are substances. Atoms are the smallest masses of the chemical elements yet known, and there are only as many kinds of atoms as of elements. To speak of an atom of a *substance*, especially of a compound substance, is a misuse of terms. In the case of elementary substances we have still to distinguish between the molecules of the substances and the atoms of the element. There is but one kind of atom of any element, but there may be several distinct elementary substances. Thus in the case of oxygen we have oxygen gas, the molecules of which consist of two atoms of oxygen, and ozone, a wholly different substance, the molecules of which consist of three atoms of oxygen. The molecules of elementary substances are formed by the aggregation of atoms of the same kind; the molecules of compound substances by the aggregation of atoms of different kinds. There are a few cases, like the vapours of mercury and zinc, where the molecule consists of a single atom. In a chemical change the molecules of the substance we call the factors break up into atoms, which group themselves into new associations to form molecules of the products.

22. Determination of Atomic Weights.

The exact determination of an atomic weight now resolves itself into a simple question of quantitative analysis. If in any process of quantitative analysis we can determine the weights of two of the *elementary substances* involved, the proportion between these quantities will be either the ratio of the atomic weights of the two elements, or else that of some simple multiple of these weights, the multiple in all cases being previously known from the relations of the compounds of the element as exhibited in such tables as we have described, or otherwise.

Ex. 74. *Atomic Weight of Zinc.*—Adapt to a small flask (100 cubic centimetres) a tightly fitting cork and exit tube leading to a pneumatic trough. Place in the flask 10 cubic centimetres of strong hydrochloric acid and 20 cubic centimetres of water. Clean scrupulously a strip of the purest sheet zinc, and accurately determine its weight, which should be as nearly 1·25 gramme as possible. Use as a receiver a glass flask of 500 cubic centimetres capacity. When all is ready, the flask filled with water standing inverted on the shelf of the trough and the mouth of the exit tube under its lip, drop the metal into the acid and quickly cork the flask. This amount of metal should yield nearly 500 cubic centimetres of hydrogen gas at the ordinary temperature of the laboratory. When the apparatus is in equilibrium notice whether any water has been sucked back towards the flask. If so, make the necessary allowance in measuring the volume of gas formed. Observe the thermometer and barometer. Sinking now the flask in the water of the trough until the level of the water is the same within and without the glass, place the palm of the hand under the mouth and quickly invert the flask and place it on the pan of the balance with the water it still holds. Take the tare, and, having exactly filled the flask with water, again weigh. The difference of these weights—that is, the weight in grammes of the water required to fill the flask, reduced for temperature if more accuracy is required—will give the volume of the gas collected at the temperature and pressure observed. Reduce the volume to standard condition, making allowance for the tension of aqueous vapour (Ex. 21 and Ex. 22). From this volume calculate the weight of hydrogen formed. Make then the proportion, As the weight of hydrogen is to the weight of zinc, so is unity (the atomic weight of hydrogen) to a value which we know, from a comparison of the zinc compounds, to be one half of the atomic weight of zinc. Double the value to find the accepted atomic weight.

In a similar way the atomic weight of magnesium may be found; and by dissolving aluminum in a solution of caustic soda the atomic weight of aluminum may be determined with great accuracy. As the atomic weights are fundamental constants in chemical calculations, it is essential that they should be known with the greatest possible precision; and hence a great deal of labor has been spent on the analytical processes used in determining their value. These processes admit of very different degrees of accuracy. There are only a very few which in the most skilful hands yield results accurate to the ten-thousandth part of the quantity estimated; and even the thousandth part is regarded as a very high degree of accuracy in chemical analysis. Most processes do not give results which can be relied upon much within one per cent; and in many cases we are forced to use methods that are far less accurate even than this. In fixing the precise value of an atomic weight our choice is usually limited, both as to the material to be analyzed and the analytical process to be used, to one or two methods; but in almost all cases there are abundant analyses of compounds of the same element which are sufficiently accurate to enable us to interpret the results obtained.

(*b*) Examples illustrating the above points may be multiplied by the teacher. Thus, the production of silver bromide from silver or the reduction of silver bromide to silver gives the means of connecting the atomic weight of bromine with that of silver. (Compare Ex. 65 and Ex 68.) So also the reduction of silver nitrate in a porcelain crucible on simple ignition will enable the student to deduce the molecular weight of silver nitrate from the atomic weight of silver, and by inference from what will soon appear he can thus determine also the molecular weight of nitric acid.

(1) It will be observed that the method of determining an atomic weight is essentially the same as the chemical method of determining molecular weights. In each case the method is based on the law of definite proportions, which applies to elementary substances as well as to compounds, only in one case the definite proportions are theoretically interpreted as the definite relative weights of atoms, and in the other as the equally definite relative weights of molecules. In all instances what we can determine with accuracy experimentally is a relative weight. What that relative weight represents is a question of interpretation. The ratio

of the two weights determined forms the first two terms of a proportion of which the third term is some known molecular or atomic weight. We thus can connect one molecular weight with another or one atomic weight with another. Moreover, since we refer both molecular and atomic weights to the same unit, we can connect a molecular weight with an atomic weight, as is constantly, in fact, done. Always our results are subject to interpretation so far as regards the question of multiple values (Ex. 73).

CHAPTER IV.

SYMBOLS AND NOMENCLATURE.

23. Chemical Symbols.

The initial letter, or letters, of its Latin name are used to represent one atom, and therefore the atomic weight, of each chemical element. Thus H stands for 1 microcrith of the element hydrogen; O, for 16 microcriths of the element oxygen; C, for 12 microcriths of the element carbon. Several atoms are represented by means of subnumerals; as S_2, which stands for $2 \times 32 = 64$ microcriths of sulphur; Cl_3, which stands for $3 \times 35·5 = 106·5$ microcriths of chlorine. Molecules are represented by writing together the symbols of the atoms of which they consist. Thus, H_2O stands for a molecule of water because each molecule of water is made up of two atoms of hydrogen and one atom of oxygen; H_2SO_4 stands for a molecule of sulphuric acid, consisting of two atoms of hydrogen, one atom of sulphur, and four atoms of oxygen; O_2 stands for a molecule of oxygen gas, an aggregate of two molecules of oxygen; while O_3 stands for a molecule of ozone gas, which, although consisting solely of oxygen, has molecules made of three atoms instead of two, and is a wholly different substance. The molecular symbol represents the molecular weight, which is the sum of the weights of the atoms of which the molecule consists. Thus the molecular weight of sulphuric acid is $(2 \times 1) + 32 + (16 \times 4) = 98$. From a molecular symbol we can always deduce the percentage composition of the substance it represents. Thus it is obvious that $\frac{2}{98}$ of a molecule of sulphuric acid consists of hydrogen, $\frac{32}{98}$ of sulphur, and $\frac{64}{98}$ of oxygen. The substance, having the same composition as the molecule, must then contain in one hundred parts—

Hydrogen	2·04
Sulphur	32·65
Oxygen	65·31
	100·00

On the principle of Avogadro, all molecular symbols represent the same volume in the state of gas; thus—

Hydrogen Gas.	Ozone.	Carbonic Dioxide.
$H_2 = 2$ m.c.	$O_3 = 48$ m.c.	$CO_2 = 44$ m.c.
Water.	Alcohol.	
$H_2O = 18$ m.c.	$C_2H_6O = 46$ m.c.,	

all represent the same gas volumes compared under the same conditions of temperature and pressure. It follows, then, that the specific gravity of a gas or vapour referred to hydrogen can be at

CHEMICAL SYMBOLS. 151

once deduced from the molecular symbol by dividing the molecular weight by 2. The weight of a litre of hydrogen when the barometer stands at 30 inches and the thermometer at 0° is 0·0896 gramme. At 273° under the same pressure the weight would be 0·0448 gramme. At 27° (a very convenient standard)* the weight would be 0·0815 gramme. Hence the weight of a litre of any gas or vapour, under either condition, may also be calculated from the molecular symbol by multiplying the specific gravity obtained as above by one of these factors.

---

* If we select 300° absolute temperature—that is, 27° C.—for our standard temperature and 30 inches of mercury for the standard pressure all reductions of gas volumes can be made with the greatest facility. A variation of one degree from this standard temperature corresponds exactly to a variation of one tenth of an inch in the barometer, and the effect of temperature can at once be eliminated by altering to a corresponding extent the height of the mercury column measuring the pressure. Moreover, as our observations are almost invariably made at temperatures below the standard (27° C., or 80·6° F.), this correction is usually additive. Assume, for example, that the temperature is 20° and the observed height of the barometer 30·3 inches, and it is desired to reduce the observed gas volume to the assumed standard. Were we to raise the temperature to 27° it is obvious that we should expand the gas; and to bring it back to its previous volume it would be necessary to increase the pressure by 0·7 inch, which corresponds, as we have stated, to seven degrees. It is thus evident that at 27° (the assumed standard) and at 31 inches the volume of gas would be the same as at 20° and 30·3 inches. The problem is then reduced to this: Given a volume of gas at 31 inches, to find what would be its volume at 30 inches; and this is obviously a very simple case under Mariotte's law. In like manner all similar problems may be solved, whether they relate to the volume of a given weight of gas or to the weight of a given volume.

The specific gravity referred to air, or the density under other conditions of temperature and pressure, may now be deduced by the methods before described. Let the student now answer, in regard to each of the molecular symbols last given, this question: What information does the symbol give in regard to the substance it represents? He ought also now to be able, without further assistance, to reverse the reasoning, and, when the percentage composition and vapour density are given, to deduce the symbol.

Assume that we have given as the result of analysis that the percentage composition of absolute alcohol is as below, and that the specific gravity of the vapour referred to hydrogen is 23. The weight of a molecule of alcohol is then 46, and the amount of each element in a molecule is that given in the second column of figures. Knowing now the weight of the several atoms, we easily find the number of each kind in one molecule—

Carbon	52·17	$24 = 2 \times 12$	C_2
Hydrogen	13·05	$6 = 6 \times 1$	H_6
Oxygen	34·78	$16 = 1 \times 16$	O
	100·00	46	

By studying the above scheme it will be seen that we shall reach the same result if we at once divide the percentages by the atomic weights, and then seek the simplest ratio of whole numbers corresponding to the results.

Since—

$$52{\cdot}17 : 13{\cdot}05 : 34{\cdot}78 = 24 : 6 : 16,$$

it must be that—

$$\frac{52{\cdot}17}{12} : \frac{13{\cdot}05}{1} : \frac{34{\cdot}78}{16} = \frac{24}{12} : \frac{6}{1} : \frac{16}{16} = 2 : 6 : 1$$

So when we do not know the molecular weight we can always find the simplest ratio of whole atoms corresponding to the percentage composition, and the true symbol of the compound must be either the symbol thus obtained or some multiple of it; and the molecular weight must be either the weight represented by this symbol or a multiple of it. In this way we can always find a symbol corresponding to the percentage composition of minerals and of similar inactive and non-volatile chemical products, and we accept the symbol thus obtained until further investigation shows that some multiple of it is more correct. In practice the result is often indefinite, because the percentages, owing to errors of analysis, are not exactly known, and we obtain a proportion which is only approximately a simple ratio of whole numbers. In this case we select what we regard as the most probable ratio, and a good deal of judgment is necessary in interpreting the results.

(1) Given the percentage composition of chloroform as follows: Carbon, 10·04; hydrogen, 0 83; chlorine, 89·13. Required the symbol, knowing that the specific gravity of chloroform vapour equals 59·75. *Ans.* $CHCl_3$.

(2) The percentage composition of sugar is given—Ex. 68 (2). What is the symbol? *Ans.* $C_{12}H_{22}O_{11}$.

(3) Calculate the percentage composition of nitro-benzol, $C_6H_5NO_2$. *Ans.* Carbon, 58·53; hydrogen, 4·07; nitrogen, 11·39; oxygen, 26·01.

(4) What is the weight of one litre of alcohol vapour at 273°?

24. Chemical Reactions.

As a chemical process consists in the breaking up of the molecules of the factors into atoms and the regrouping of the same atoms without loss to form the molecules of the products, it is obvious that every chemical change may be represented by an equation, writing the symbols of the molecules of the factors in the first member and the symbols of the molecules of the products in the second member, using figures like coefficients in algebra to indicate the number of molecules involved in the process and plus signs to separate the symbols. Thus the chemical change in the preparation of oxygen gas from potassic chlorate (Ex. 23), is represented by the equation—

$$\underset{\text{Potassic Chlorate.}}{2\,KClO_3} = \underset{\text{Potassic Chloride.}}{2\,KCl} + \underset{\text{Oxygen Gas.}}{3\,O_2}$$

Such an expression is called in chemistry a reaction. It presupposes a knowledge of the symbols of the products and of the factors and also of the number of molecules which concur in the process. Again, the preparation of hydrogen gas from zinc and dilute sulphuric acid (**Ex. 25**) is represented thus:

$$\underset{\text{Zinc.}}{Zn} + \underset{\text{Sulphuric Acid.}}{\underset{\text{Dilute}}{(H_2SO_4 + Aq)}} = \underset{\text{Zinc Sulphate.}}{\underset{\text{Solution of}}{(ZnSO_4 + Aq)}} + \underset{\text{Gas.}}{\underset{\text{Hydrogen}}{H_2}}$$

Here Aq indicates an indefinite amount of water used to dilute the acid or dissolve the salt. So also the formation of ammonia (**Ex. 45**) is expressed by this reaction:

$$\underset{\text{Nitric Oxide.}}{2\,NO} + \underset{\text{Hydrogen Gas.}}{5\,H_2} = \underset{\text{Water.}}{2\,H_2O} + \underset{\text{Ammonic Gas.}}{2\,NH_3}$$

In like manner the student should review all the experiments he has tried, and with the aid of the teacher learn to write the reaction in every case. He should be required to state the full significance of every reaction he writes until he has acquired a complete mastery of this symbolical language.

If correctly written, a chemical reaction illustrates always the first two, and, when it involves aëriform factors or products, all three of the fundamental laws of chemistry. The law of conservation of mass is expressed by the equation sign, the law of definite proportions by weight is indicated by the definite numerical values of the several terms, and the law of definite proportions by volume is seen in the simple ratios of the coefficients. In like manner the system of chemical symbols involves most of the principles which have been discussed in this course of experiments. For example, we see at once from the reaction why in determining the atomic weight of zinc in Ex. 74 the observed value was doubled, since what we observed was the ratio $H_2 : Zn$, and what we required the ratio $H : Zn$. So also in Ex. 66 (*b*) we see that water must have been formed by the reactions in question. Indeed, so fully do the symbols embody the general principles of chemistry that students are apt to infer that these principles have been deduced from the symbols, just as in mathematics similar general results have been discovered by the working out and interpretation of algebraic formulæ; and when, as in many text books, the phenomena are subordinated to the symbolical expression, this false impression is inevitable. Chemistry is an inductive, not a deductive science, like mathematics, and chemical symbols differ essentially from mathematical formulæ. In mathematics everything that can be legitimately deduced from an algebraic equation must be true; but this is far from being the case with chemical reactions. Chemical symbols simply stand for the facts and theories they were devised to express and for nothing more. To secure the peculiar discipline of the physical sciences it is essential that they should be studied as they have been built up. The student must begin by observing the phenomena, and be led up to the general principles through his own inferences, and this is the order which has been followed in preparing this book. To begin with an abstract statement of these principles, or, what amounts to the same thing, to express at once every phenomenon observed in symbolical language which embodies these principles, is to invert the natural order and to abandon the inductive method. Nevertheless, such are the perfection and grasp of this system of symbols that it is of the greatest value in aiding us to realize relations and foresee results which without it we might not have discovered.

(1) The direction above given, that the student should review all the experiments heretofore described and write all the reactions of the processes described so far as it is possible, can not be too strongly insisted upon. Without such practice the student can not be expected to grasp the subject; but with it many of the relations before observed will become clear and the facts will all appear in a clearer light. To aid him we give below the more important reactions, the figure prefixed being the number of the experiment under which the process is described or indicated:

14. (a) $(H_2,C_4H_4O_6 + HNaCO_3 + Aq) = (HNa,C_4H_4O_6 + H_2O + Aq) + CO_2$
14. (b) $(BaCl_2 + Na_2SO_4 + Aq) = BaSO_4 + (2NaCl + Aq)$
16. $2H_2O = 2H_2 + O_2$
23. $2KClO_3 = 2KCl + 3O_2$
24. (a) $P_4 + 5O_2 = 2P_2O_5$
24. (c) $C + O_2 = CO_2$

25. $Zn + (H_2SO_4 + Aq) = (ZnSO_4 + Aq) + H_2$
27. $2H_2 + O_2 = 2H_2O$
32. $S + O_2 = SO_2$
     $(SO_2 + H_2O + Aq) = (H_2SO_3 + Aq)$
33. $2SO_2 + O_2 = 2SO_3$
     $(SO_3 + H_2O + Aq) = (H_2SO_4 + Aq)$
     $(BaCl_2 + H_2SO_4 + Aq) = Ba_2SO_4 + (2HCl + Aq)$
     $(P_2O_5 + 3H_2O + Aq) = (2H_3PO_4 + Aq)$
34. $(NaCl + H_2SO_4 + Aq) = (HNaSO_4 + Aq) + 2HCl$
35. $2HCl + Na_2 = 2NaCl + H_2$
36. $(MnO_2 + 4HCl + Aq) = (MnCl_2 + 2H_2O + Aq) + Cl_2$
     $H_2 + Cl_2 = 2HCl$
39. (a) $(CaCO_3 + 2HCl + Aq) = (CaCl_2 + H_2O + Aq) + CO_2$
39. (b) $(CaO_2H_2 + Aq) + CO_2 = (CaCO_3 + (H_2O + Aq)$
40. (a) $CO_2 + C = 2CO$
41. $C_2H_6O + H_2SO_4 = (H_2SO_4 . H_2O) + C_2H_4$
     $CH_4 =$ Marsh Gas, $C_2H_4 =$ Ethylene, $C_2H_2 =$ Acetylene.
43. (a) $KNO_3 + H_2SO_4 = HKSO_4 + HNO_3$
43. (b) $2HNO_3 + S = H_2SO_4 + 2NO$
43. (c) $2HNO_3 + 5Cu = 5CuO + H_2O + N_2$
44. (a) $3Cu + (8HNO_3 + Aq) =$
     $\qquad\qquad\qquad (3CuN_2O_6 + 4H_2O + Aq) + 2NO$
     $2NO + O_2 = 2NO_2$
44. (b) $10NO + P_4 = 2P_2O_5 + 5N_2$
45. $2NO + 5H_2 = 2H_2O + 2NH_3$
48. $2Mg + O_2 = 2MgO$
     $MgO + H_2O = MgO_2H_2$
     $MgO + (H_2SO_4 + Aq) = (MgSO_4 + H_2O + Aq)$
     $MgO_2H_2 + (H_2SO_4 + Aq) = (MgSO_4 + 2H_2O + Aq)$
     $Mg + (H_2SO_4 + Aq) = (MgSO_4 + Aq) + H_2$
     $2Zn + O_2 = 2ZnO$
     ZnO does not unite directly with water.
     $ZnO + (H_2SO_4 + Aq) = (ZnSO_4 + H_2O + Aq)$
49. $Na_2 + (2H_2O + Aq) = (2NaOH + Aq) + H_2$
     $(NaOH + HCl + Aq) = (NaCl + H_2O + Aq)$
     $(NaOH + HNO_3 + Aq) = (NaNO_3 + H_2O + Aq)$
     $(NaOH + CO_2 + Aq) = (HNaCO_3 + Aq)$
50. (c) $Na_2 + Cl_2 = 2NaCl$
     $2Na_2 + O_2 = 2Na_2O$

CHEMICAL REACTIONS.

$$Na_2O + H_2O = 2NaOH$$
$$(Na_2O + 2HCl + Aq) = (2NaCl + H_2O + Aq)$$
$$(NaOH + HCl + Aq) = (NaCl + H_2O + Aq)$$

51. $(2NH_3 + H_2SO_4 + Aq) = ((NH_4)_2SO_4 + Aq)$
$(NH_3 + HNO_3 + Aq) = (NH_4NO_3 + Aq)$
$(NH_3 + HCl + Aq) = (NH_4Cl + Aq)$
$(2NaOH + H_2SO_4 + Aq) = (Na_2SO_4 + 2H_2O + Aq)$
$(NaOH + HNO_3 + Aq) = (NaNO_3 + H_2O + Aq)$
$(NaOH + HCl + Aq) = (NaCl + H_2O + Aq)$

53. (a) $Cu + O = CuO$
$CuO + H_2 = Cu + H_2O$

54. (a) $(CuO + H_2SO_4 + Aq) = (CuSO_4 + H_2O + Aq)$
$(CuO + 2HNO_3 + Aq) = (CuN_2O_6 + H_2O + Aq)$
$(CuO + 2HCl + Aq) = (CuCl_2 + H_2O + Aq)$

57. (a) $Fe + H_2SO_4 + Aq) = (FeSO_4 + Aq) + H_2$

58. FeO = Ferrous Oxide
Fe_2O_3 = Ferric Oxide
$FeOSO_3$ (or $FeSO_4$) = Ferrous Sulphate
$Fe_2O_3.3SO_3$ = Ferric Sulphate

59. (a) $Fe + S = FeS$

59. (b) $FeS + (H_2SO_4 + Aq) = (FeSO_4 + Aq) + H_2S$
$FeO + (H_2SO_4 + Aq) = (FeSO_4 + Aq + H_2O)$

65. $(AgNO_3 + HCl + Aq) = AgCl + (HNO_3 + Aq)$
$2AgCl + H_2 = 2Ag + 2HCl$

68. $(AgNO_3 + KBr + Aq) = (KNO_3 + Aq) + AgBr$

73. $(MnO_2 + H_2SO_4 + H_2C_2O_4 + Aq) =$
$(MnSO_4 + 2H_2O + Aq) + 2CO_2$

(2) To ensure a full comprehension of the subject the teacher should ask such questions, as the following:

Does it appear from 27 that when hydrogen unites with oxygen two volumes of the first combine with one volume of the second to form two volumes?

Does it appear that when either charcoal or sulphur burn in oxygen gas the volume of the product is the same as the volume of the oxygen consumed?

In the ordinary process of preparing oxygen gas, how does the residual salt differ in composition from the salt used?

Both zinc and zinc oxide when dissolved in dilute sulphuric acid yield the same zinc sulphate. Why is hydrogen gas evolved in the first case and not in the second?

How do you explain the production of nitric acid from nitre—Ex. 43 (*a*)?

When iron dissolves in nitric acid either nitric oxide or nitrogen gas is evolved (as in case of copper), while from dilute sulphuric acid the same metal liberates hydrogen. Why the difference?

When to a solution of a silver coin we add hydrochloric acid all the silver is precipitated as chloride, but none of the copper. Why this selection?

By the judicious use of such questions the student will be led to think, the deadening effect of mechanical routine will be avoided, and the teaching power of the course greatly increased.

25. Stochiometry.

Since in writing a chemical reaction the relative weights of all the factors and products are necessarily implied, it follows that if the total weight of any one substance concerned in the process is given the weight of every other may be calculated. It is only necessary to make the proportion—

As the total molecular weight of the given substance is to the total molecular weight of the required substance, so is the gross weight given to the gross weight required.

By total molecular weight is here meant the simple molecular weight of the substance multiplied by the coefficient with which it appears in the reaction. If in such problems a volume is given this volume must be reduced to weight by the simple method already described before applying the rule; and when the volume of a factor or product is sought the reverse reduction is readily made after the weight is known. The relation between any two gas volumes is of course directly seen on inspecting the reaction, and needs no calculation. The student ought to have a great deal of practice in stochiometrical calculation. A very large number of problems of this sort will be found in the author's Chemical Philosophy, and there are many works wholly devoted to the subject. The teacher, however, will add interest to a necessarily dry subject if he constructs problems of his own, based on the experiments which the student has actually performed.

(1) How many grammes of common salt can be made from 25 grammes of sodium bicarbonate?

(2) In the process of making nitric acid, how many grammes of sulphuric acid will be required to every kilogramme of nitre, assuming that the acid used contains 95 per cent of H_2SO_4? How many cubic centimetres would be required? (Such an acid has the specific gravity 1·84.)

(3) How many grammes of charcoal and how many of sulphur will burn in one litre of oxygen measured at standard conditions? What will be the weight of one litre of each of the products under the same conditions?

(4) How many grammes of potassium chlorate are required to yield four litres of dry oxygen (standard conditions)?

(5) When dissolved in acid 1·25 gramme of zinc will yield what volume of hydrogen gas collected over water

when temperature of room is 20 and barometer stands at 750 millimetres ?

(6) In preparing ammonia gas from nitric oxide and hydrogen gas in what proportions by volume should the last two be mixed ?

(7) In preparing nitric oxide, how many grammes of copper will be required to each litre of gas if the product is only nitric oxide ? How many if the product is wholly nitrogen gas ?

(8) A cubic decimetre of marble contains how many times its own volume of carbonic-dioxide gas ? Specific gravity of marble, 2·75.

(9) The reactions involved in these problems are all given in the list under the preceding division   The student should not limit his study to the few problems here given as examples, and it must be borne in mind that the same principles apply to any chemical process, however complex or however so many simple reactions it may involve, provided only that the whole material from one reaction passes forward to the next.

26. Nomenclature.

Before the present century the names given to chemical products were almost wholly arbitrary, and a few of these, like oil of vitriol, blue vitriol, sugar of lead, calomel, and Epsom salts, still remain in common use. In 1787 a systematic nomenclature was devised by a committee of the French Academy of Sciences, under the lead of Lavoisier, in which the name of a substance was made to indicate its composition, and at the time of its adoption and for more than fifty years afterwards it was probably the most perfect nomenclature which any science ever enjoyed. It was based, however, on the dualistic theory of Lavoisier, and when the science outgrew the theory the old names lost much of their significance and appropriateness. Nevertheless, the main features of the Lavoisierian nomenclature are still preserved, although with some variations of usage as to details and the introduction of many arbitrary names, like carbinol, phenol, pinakone, demanded by the necessities of a rapidly expanding science. The old nomenclature is so out of harmony with our modern conceptions that it would be impossible to explain its full significance without entering into details which would be out of place in an elementary course. A few of the rules should be stated by the teacher and the use of the ordinary terminations and prefixes so far explained as to render the usually occurring names intelligible. All that is required may be found in any elementary text book on chemistry.

CHAPTER V.

MOLECULAR STRUCTURE.

27. Quantivalence.

OF all chemical reactions by far the most common is a class in which, judging from the products, the only change that takes place is an interchange of atoms or groups of atoms between two sets of molecules, leaving all relations otherwise the same as before. Such reactions are described as metathetical, and the process is termed metathesis. Our chemical symbols here come to our aid by enabling us to form a clear idea of what is meant by these terms. Thus in the reaction of a solution of silver nitrate or a solution of potassium bromide (Ex. 68)—

$$(AgNO_3 + KBr + Aq) = (KNO_3 + Aq) + AgBr$$

it is obvious that Ag changes place with the K. So also in the reaction by which hydrogen gas is made from zinc and dilute sulphuric acid—

$$(H_2SO_4 + Aq) + Zn = (ZnSO_4 + Aq) + H_2$$

it is equally obvious that Zn changes place with

H_2. Again in the ordinary test for sulphuric acid—

$$(H_2SO_4 + BaCl_2 + Aq) = BaSO_4 + (2\ HCl + Aq)$$

it is evident that Ba has changed place with H_2. The following experiment will further illustrate this point:

Ex. 75. *Metathesis.*—Pour ten cubic centimetres of water into a test tube and dissolve in it one gramme of silver nitrate. Immerse in the liquid a small strip of pure copper, whose weight must be accurately determined and must not exceed eighteen centigrammes. After the silver has separated, wash the powder on to a small filter and continue to pour water on to the filter until it runs through tasteless. Dry the filter on the tunnel. Remove when dry, and after carefully wrapping the loose paper round the silver powder place the ball in a tared porcelain crucible and slowly heat to redness until the paper has been burned, when the silver will appear bright. When cold again weigh the crucible and find the weight of the silver, which should be to the weight of the copper approximately as $Ag_2 = 216 : Cu = 63 \cdot 6$.

Immerse now in the blue liquid decanted from the silver a strip of zinc. The copper which had passed into solution in the previous process will now be precipitated. The reactions may be written:

$$Cu + (2\ AgNO_3 + Aq) = Ag_2 + (Cu(NO_3)_2 + Aq).$$
$$Zn + (Cu(NO_3)_2 + Aq) = Cu + (Zn(NO_3)_2 + Aq).$$

Obviously, in the first, one atom of copper replaces two atoms of silver, and in the second one atom of zinc replaces one atom of copper; and in studying these reactions it must be remembered that the symbols correctly represent the atomic relations. Now, by studying in a similar way a very large number of metathetical reactions, it appears that the atoms of hydrogen, lithium, sodium, potassium, cæsium, rubidium, silver, thalium, chlorine, bromine, and iodine are alike in this, that while among themselves they can be exchanged atom for atom, they are replaced by all other atoms in groups of two, three, four, or more. The atoms enumerated appear to have the smallest exchangeable value and are said to be univalent, while atoms which will fill the place of two univalent atoms are said to be bivalent, those that can fill the place of three trivalent, those that can fill the place of four quadrivalent, etc. So also the terms quantivalence and multivalence. The facts here stated suggest at once the conception of molecular structure, for it would seem as if the parts of a molecule must be bound together in definite relative positions in order to render such substitutions possible. This conception of structure is greatly widened and strengthened when we compare together the symbols of molecules formed by the union of atoms having different degrees of quantivalence, as shown in metathetical reactions similar to those just described, for it appears that the combining power corresponds exactly to the replacing power. In making such comparisons it must constantly be borne in mind that the symbol represents in every case our knowledge of the composition and relations of the substance, and that the symbolical language thus enables us to bring before the mind in one view the results of long-continued and laborious investigation. We give below the symbols of four well-known and typical compounds of hydrogen:

Hydrochloric Acid.	Water.	Ammonia Gas.	Marsh Gas.
HCl	H_2O	H_3N	H_4C

In these compounds the atoms Cl, O, N, and C are united with the same number of univalent atoms, which, under other circumstances, they might replace. So, also, we have—

Common Salt.	Baric Chloride.	Antimonious Chloride.	Zirconic Chloride.
$NaCl$	$BaCl_2$	$SbCl_3$	$ZrCl_4$

Compare now with these the corresponding oxides—

$$Na_2O \quad BaO \quad Sb_2O_3 \quad ZrO_2,$$

and it will be seen how we are led to the conclusion that in these molecules the atoms of lower quantivalence are united to those of higher quantivalence, which are, as it were, the nucleus of the molecule, and serve like a clamp to bind the parts together. In order to express this we often write the symbols as below, using dashes to indicate what we call the bonds. Symbols thus written are said to be *graphic*, while as before written they are not inappropriately spoken of as *empirical*.

$$\text{H-Cl} \qquad \text{H-O-H} \qquad \overset{\text{H}}{\underset{}{\text{H-N-H}}} \qquad \overset{\text{H}}{\underset{\text{H}}{\text{H-C-H}}}$$

$$\text{Na-Cl} \qquad \text{Cl-Ba-Cl} \qquad \overset{\text{Cl}}{\underset{}{\text{Cl-Sb-Cl}}} \qquad \overset{\text{Cl}}{\underset{\text{Cl}}{\text{Cl-Zr-Cl}}}$$

$$\text{Na-O-Na} \qquad \text{Ba-O} \qquad \text{O-Sb-O-Sb-O} \qquad \text{O=Zr=O}$$

This is but a very short step in our reasoning, and yet it opens to view at once a very definite structure. Notice that the only inference we have drawn is that the atoms, instead of being indiscriminately piled together, are united each separately to the multivalent atom or atoms of the group. This inference granted, we can take another step.

When sodium acts on water we have a simple metathesis—

$$(2\,\text{H-O-H} + \text{Aq}) + \text{Na}_2 = (2\,\text{Na-O-H} + \text{Aq}) + \text{H}_2,$$

and the product Na–O–H must have the same structure as H–O–H, containing only Na in place of one of the hydrogen atoms. In a similar way, when magnesium acts on water it must be that—

$$\begin{matrix}H\text{-}O\text{-}H\\H\text{-}O\text{-}H\end{matrix} + Mg = Mg\begin{matrix}\text{-}O\text{-}H\\\text{-}O\text{-}H\end{matrix} + H_2.$$

And here it is evident that the multivalent atom Mg binds together two molecules of water, and thus we can conceive how complex molecules may be built up, and we also see that the replacing and combining powers are merely different manifestations of the same atribute of the atoms which we express by the term "bonds"; and hence the reason that the two powers necessarily correspond.

The products whose formation and structure we have studied are two members of a very large class of substances, all having similar relations, and which must have a similar structure. The symbols of four other members of the same class are given on the next line:

$$K\text{-}O\text{-}H \quad Ca\begin{matrix}\text{-}O\text{-}H\\\text{-}O\text{-}H\end{matrix} \quad Zr\begin{matrix}\text{-}O\text{-}H\\\text{-}O\text{-}H\\\text{-}O\text{-}H\end{matrix} \quad Al\begin{matrix}\text{-}O\text{-}H\\\text{-}O\text{-}H\\\text{-}O\text{-}H\\\text{-}O\text{-}H\\\text{-}O\text{-}H\\\text{-}O\text{-}H\end{matrix}$$

Bodies of this class are called hydrates, and the chief feature in their structure is that they contain atoms of hydrogen united with a multivalent nucleus through atoms of oxygen; and, further, the one chemical relation which marks all such substances is that the atoms of hydrogen so united are easily replaced by simple metathetical reactions.

Ex. 76. *Replacement of Hydrogen in Sodic Hydrate.*—In a small flask (50 cubic centimetres) dissolve 5 grammes of caustic soda in 10 cubic centimetres of water; add a strip of aluminum weighing less than three decigrammes; connect with pneumatic trough, and collect the hydrogen gas evolved. Measure the gas volume, observing thermometer and barometer; correct for tension of aqueous vapor and calculate the weight of hydrogen obtained. Compare this weight with the weight of aluminum, and estimate how many atoms of hydrogen must have been replaced by each double atom of aluminum (54 microcriths) indicated by the barred symbol.

If now, in this connection, the student will review the familiar experiment by which hydrogen gas is usually made, he will see that this process is also a metathetical reaction in which hydrogen atoms have been replaced by atoms of zinc; and if we take such reactions as an indication of the type of structure we have assigned to the hydrates, it will appear on further study that most of the active agents of chemistry, including both acids and alkalies, must be classed in the same group. To class acids and alkalies in the same group of compounds seems at first sight very anomalous, for in most respects their properties are the direct opposites each of the other. Nevertheless, not only do these bodies resemble each other in the one essential relation of a hydrate, but also they may be grouped in series varying so gradually from strong alkalies at one end to strong acids at the other that no natural dividing line can be found. Moreover, the distinction between an acid and a base is of a relative, and not of an absolute character, as is shown by the fact that in such series as have been mentioned a given member may act as an acid towards the members at one end of the list, and as a base towards those at the other end. If now we study the symbols of the ordinary acids, and arrange the hydrogen atoms after the pattern of a hydrate, we shall get such a result as this:

Nitric Acid.

$H-O-NO_2$

Sulphuric Acid.

$\begin{matrix} H-O- \\ H-O- \end{matrix} SO_2$

Phosphoric Acid.

$\begin{matrix} H-O \\ H-O \\ H-O \end{matrix} PO$

And it will be seen that though here, as before, the hydrogen atoms are united through oxygen atoms to a multivalent atom which serves as the nucleus of the molecule, this atom of high quantivalence acting as an atomic clamp is one of a group of atoms. Such groups are termed in chemistry compound radicals, and the chain H-O- is called hydroxyl. If the reasoning has been followed thus far the student will be prepared to admit that the relations of acids and bases which play such an important part in the science of chemistry depend upon molecular structure. But why the opposition between these bodies? As otherwise they have the same structure, it is evident that the antagonism must be connected with the atoms or compound radical with which the hydroxyl groups are associated; and when we compare the nuclei of the respective molecules we find a very manifest difference between the nucleus of a marked alkali and the nucleus of a pronounced acid. In the alkali the nucleus is a metallic atom, like sodium or potassium; in the acid it is a non-metallic atom, or a group of such atoms, like nitrogen or sulphur. Moreover, we find that while it is very easy to replace the hydrogen atoms either of an alkali or of an acid by atoms unlike the nucleus, it is difficult to replace them by atoms similar to the nucleus. Thus it is difficult to change Na–O–H to Na–O–Na, but very easy to change it to Na–O–Cl or Na–O–NO$_2$. The phenomena point to a polarity in the molecule similar to that of a magnet, and only by such analogies can we explain them.

The objects of this discussion have been, in the first place, to give an idea of the mode of reasoning by which our knowledge of molecular structure is reached; and, in the second place, to explain the distinction between acids, alkalies, and salts. This distinction, however much it may be described, can not be made clear except through the molecular structure on which it is supposed to depend. Towards the first object we have been able to advance only a very few steps, but far enough to point out the way by which the complex structure of organic compounds has been unravelled and the whole subject of organic chemistry developed. As towards the second object, we hope we have been able to make clear that acids and alkalies have in their larger relations similar qualities and a similar structure, and that they differ in the character of the nuclei to which the hydroxyl groups are united; and, further, that when the hydrogen atoms thus united are replaced, in compounds of either class, by atoms opposite in qualities to the atoms of the nuclei, the products are salts—so called because for the most part they are bodies that can be readily crystallized. In this connection there are one or two other points to be noticed before leaving the subject.

As has been shown, an acid most readily combines with an alkali to form a salt; and the reason seems to be that the polarity of the molecules tends to bring atoms of opposite characters to the two ends, and determines a metathesis thus—

$$Na-O-H + H-O-NO_3 = Na-O-NO_3 + H-O-H$$
$$Ca=O_2=H_2 + H_2=O_2=SO_2 = Ca=O_2=SO_2 + 2H-O-H$$

The number of hydroxyl groups in a hydrate, whether acid or alkali, measures what we call its atomicity. Thus sodic hydrate is monatomic, and aluminic hydrate hexatomic. In an acid the atomicity is frequently called basicity, and just as the metallic atoms associated with the acid nuclei in a salt are often spoken of as basic radicals, so the metallic hydrates used for neutralizing the acid, as in the above reaction, are often termed bases, or the base of the salt.* Thus nitric acid is monobasic, sulphuric acid is dibasic, and phosphoric acid is tribasic. Hence, while nitric acid will form only one salt with sodium, sulphuric acid will form two, and phosphoric acid will form three. This point is illustrated by the following symbols—

$$Na-O-NO_2$$

$$\begin{matrix} H-O \\ Na-O \end{matrix} SO_2 \qquad \begin{matrix} Na-O \\ Na-O \end{matrix} SO_2$$

$$\begin{matrix} H-O \\ H-O \\ Na-O \end{matrix} PO \qquad \begin{matrix} H-O \\ Na-O \\ Na-O \end{matrix} PO \qquad \begin{matrix} Na-O \\ Na-O \\ Na-O \end{matrix} PO$$

And the fact that the several salts which the symbols show to be possible can be prepared is in harmony with the molecular structure we have described.

The chief features of the type of molecular

---

* The term "base" is used in a broader sense than the word "alkali," and is applied to any hydrate which will unite with an acid..

structure we have unfolded are very strikingly illustrated by the formation and decomposition of ammonium nitrate. The molecules of ammonia gas must have (as we have seen) the simple structure—

$$\overset{\overset{\text{H}}{|}}{\text{H}-\text{N}-\text{H}};$$

but when the gas dissolves in water the solution acquires properties so closely resembling those of a solution of sodium hydrate that we naturally conclude that new molecules have been formed by combination with water having a structure similar to $\text{Na}-\text{O}-\text{H}$; that is—

$$\text{H}_3\text{N} + \text{H}-\text{O}-\text{H} = \text{H}_4\text{N}-\text{O}-\text{H}.$$

When next we neutralize ammonium hydrate with nitric acid, as in Ex. 46 (*a*), the reaction must be like the reaction of the same acid on sodium nitrate given above, or—

$$\text{H}_4\text{N}-\text{O}-\text{H} + \text{H}-\text{O}-\text{NO}_2 = \text{H}_4\text{N}-\text{O}-\text{NO}_2 + \text{H}-\text{O}-\text{H}.$$

Here, then, if our reasoning is correct, we have the molecule of a salt having as a nucleus $\text{N}-\text{O}-\text{N}$, but with hydrogen atoms attached to the nitrogen atom which forms one pole of the molecule, and oxygen atoms united to the nitrogen atom which forms the opposite pole; and that this is the true structure is indicated by the fact that when we simply heat the salt, the oxygen and hydrogen atoms, thus for a time kept apart, rush into combination, and form molecules of water, leaving the nuclei free to become the molecules of a well-known substance called nitrous-oxide gas.

Ex. 77. *Preparation of Nitrous Oxide.*—Connect a small flask (fifty cubic centimetres) by means of perforated corks and glass tubes, first with a test tube and second with a pneumatic trough. Place in the flask twenty-five grammes of ammonium nitrate, and mount the apparatus so that while the flask is held by a retort holder the test tube may stand in a beaker of water and the exit tube may open under the mouth of a glass jar standing full of water and inverted on the shelf of the trough. Cautiously heat the salt until it melts, and then press the heat until decomposition ensues. Water will distil over and collect in the test tube, while nitrous oxide will bubble up and displace the water in the jar. When the jar is filled, seal it and preserve the gas for comparison.

Ex. 78. *Composition of Nitrous Oxide.*—Into a jar of nitric oxide, prepared as in Ex. 44 (*a*), cautiously pour 100 cubic centimetres of a concentrated solution of green vitriol acidified with hydrochloric acid, seal the jar, and shake the solution with the gas so long as absorption continues. Open now the mouth of the jar under water, and, after comparing the residual gas volume with the original volume of the nitric oxide, identify the product as the same substance which was formed in the last experiment. Knowing that the symbol of nitric oxide is NO, and that the effect of the green vitriol is to withdraw oxygen, what inference can you make in regard to the composition of nitrous oxide and as to the nature of the molecular change which has taken place in this experiment.

(1) The molecular structure of ammonium nitrate thus developed may serve to give some conception of the conditions to which modern explosives, like nitro-glycerin and gun cotton, owe their remarkable relations. In the molecules of these explosives it is supposed that three or more of the nuclei N – O – N are bound together by groups of hydrogen and carbon atoms at one end of a multiple chain, while oxygen atoms, in sufficient numbers to unite with all the carbon and hydrogen, are attached at the other end. By this structure the intensely powerful affinities between the great fire element and the combustibles are for a time held in abeyance. But when the equilibrium is disturbed the atoms thus held apart rush together and a great volume of aëriform products are suddenly developed whose enormous expansive force produces the destructive effects so well known.

CHAPTER VI.

THERMAL RELATIONS.

28. Heat of Chemical Action.

EVEN the most elementary course, proposing to treat only of the fundamental principles of chemistry, would be incomplete without some discussion of the thermal relations of chemical changes, and the most striking chemical experiments will have to the student no more meaning than fireworks if they remain in his mind as merely brilliant phenomena. After what must be here assumed to have been already learned, the student should be prepared to understand the treatment of this subject in the last chapters of the author's New Chemistry, and more fully in the chapter on the "Thermal Relations of Atoms" in his Chemical Philosophy. As an introduction to the subject, the student should try the following experiments, using for the purpose the calorimeter, already fully described (Ex. 8), but when corrosive liquids are used substituting for the inner brass vessel as thin a beaker glass as can be had, of about the same capacity, and filling the space between the glass and the sides of the chamber with layers of wool wadding, caught together so that the beaker can readily be removed and replaced. For accurate experiments a dish made of thin platinum plate is always to be preferred. In using a beaker the heat absorbed by the glass becomes a quantity of importance. The glass must therefore be weighed; and the weight of the glass multiplied by the specific heat of glass (Ex. 8 (4)) gives a value which is called the thermal water equivalent, and this in every experiment is to be added to the weight of the water.

Ex. 79. *Heat of Hydration and of Solution.*—Place in the calorimeter about 300 grammes of water, weighing the amount accurately to a gramme. Prepare and pulverize 35 grammes of anhydrous sodic sulphate by driving off the water from the crystallized salt (Glauber's salts). Keep the salt between watch glasses over the beaker of water and under the cover of the calorimeter until a perfect equilibrium of temperature is reached. Then stir the salt into the water with a glass rod, and observe the rise of temperature. Calculate the number of units of heat evolved by 142 grammes of the salt.* This number is the molecu-

---

* In making this and similar calculations it must be remembered that is the whole mass of the solution, and not the water merely, whose temperature is changed. To obtain strictly accurate

lar weight in gramme units; and it is convenient to state results on this basis, as we can then carry our calculations through successive reactions without constant reduction. Repeat now the same experiment, but use 79·3 grammes Glauber's salts ($Na_2SO_4 . 10H_2O$) in fine crystals. Calculate as before for one molecule in grammes, and compare the two results. How much heat is liberated in the union of Na_2SO_4 with $10H_2O$?

Ex. 80. *Heat of Neutralization.*—Weigh in a glass-stoppered vial about twenty-five grammes of strong sulphuric acid, the specific gravity of which has been previously accurately ascertained. Mix this acid with about 250 cubic centimetres of water and give time to cool before pouring into the calorimeter. Finding from the tables the amount of H_2SO_4 thus taken, calculate the amount of Na–O–H required to neutralize the acid and weigh out about one fifth more than the calculated amount in order to insure an excess. Dissolve the alkali also in about 250 cubic centimetres of water and place the vessels holding the two solutions under cover in a protected place at the side of the cal-

---

results, we should know the specific heat of the solution; but for all practical purposes it is sufficiently accurate to count one cubic centimeter of the solution, measured at 4° C., as the thermal equivalent of one gramme of water. The simplest way is, after the close of the experiment, to weigh the solution and determine its specific gravity with a delicate spindle hydrometer graduated at 4° C. Then the weight, divided by the specific gravity, is the thermal water equivalent in grammes.

orimeter. When both solutions have cooled to the same temperature, pour them together into the calorimeter and observe the rise of temperature. Calculate for one molecule H_2SO_4 in grammes.

Ex. 81. *Heat of Chemical Action.*—Mix 250 cubic centimetres of water with fifty cubic centimetres of strong sulphuric acid, and when the mixture is cold place it in the calorimeter. Scrupulously clean a strip of sheet zinc about two inches wide by five inches long. Accurately weigh the zinc. Plunge the strip into the acid and allow the action to continue until the temperature has risen three or four degrees. Then remove the metal, note the rise of temperature, and after washing strip with water and alcohol, dry, and determine the loss of weight. Calculate the amount of heat evolved for each molecule in grammes of $ZnSO_4$ formed.

Ex. 82. *Heat of Precipitation.* — Dissolve a weighed amount—about twenty grammes—of crystallized baric chloride ($BaCl_2 . 2 H_2O$) in 250 cubic centimetres of water. Calculate the quantity of sulphuric acid of known specific gravity required to decompose the salt, and mix the acid in slight excess of the calculated amount with 250 cubic centimetres of water. Handle the solutions as in Ex. 80, pouring first the acid and afterwards the salt solution, slowly and with constant stirring, into the calorimeter. Calculate for each molecule of BaSO₄ (in gramme units) formed.

Ex. 83. *Crystallized Cupric Sulphate.*—Dissolve a weighed amount—about twenty grammes —of blue vitriol in about 250 cubic centimetres of water and place the solution in the calorimeter. Plunge in the solution a strip of zinc prepared as in Ex. 81 and stir until the copper is wholly precipitated, and then note the rise of temperature. Calculate for every 63·6 or every atom in grammes of copper reduced.

Unfortunately the fundamental experiments on this subject—such, for example, as those on the heats of combustion of hydrogen, carbon, and sulphur, or the heat of formation of hydrochloric acid—are out of the reach of elementary students, and indeed of most teachers, but the general principles involved can be readily made clear. The chief points to be insisted on are:

(1) It follows from the principle of conservation of energy, and has been fully proved by investigation, that in a series of chemical changes the total amount of heat developed depends wholly on the initial and final states of the system, and is not dependent on the intermediate steps. Thus ninety-six grammes of sulphuric acid consists of thirty-two grammes of sulphur, sixty-four grammes of oxygen, and two grammes of hydrogen, and can be prepared by combining these relative amounts of roll brimstone, oxygen gas, and hydrogen gas in several different ways. But whatever may be the series of processes employed, the total amount of heat evolved in the production of ninety-eight grammes of this definite compound from the several elementary sub-

stances will be 193,100 units. Hence, conversely, if by a series of analytical processes we resolve back this compound into the same elementary substances in the same condition an equal amount of heat will be absorbed. It is only exceptionally the case that we can prepare compounds by the direct union of elementary substances, and even when we can it is rarely that we can measure the heat thus developed; but the above principle renders this unnecessary. We can always determine the heat evolved in the production of a compound from elementary substances (or conversely) by measuring the heat evolved at each step of the successive operations by which it may be made, and we are thus able to choose such processes as are adapted to thermal measurements; and in the investigations of thermo-chemistry a great deal of ingenuity has been shown in this selection or in devising new processes which are compatible with the methods of calorimetry. In this manner what is termed the heat of formation of a large part of known compounds has been measured. In making our calculations and in stating results we adopt the system already referred to (Ex. 79), and a table giving the more important data will be found in the author's work on Chemical Philosophy.

The thermal relations have led to the division of compounds into two large classes—*exothermous bodies* (by far the larger class), whose formation from *known elementary substances in their familiar state* is attended with the evolution of heat, and *endothermous compounds*, of which the reverse is true.

It also follows from the principle we have been considering that if we begin with the same substance in the same state and by different processes reach two different products, the difference in the heat evolved in the two cases is that required to pass from one product to the other, or, what is an obvious corollary, if the initial states are different and the final results in all respects the same, then the same relation will hold between the initial states. This deduction gives us a very simple means of determining the heat of combination in a great number of cases where direct union is impossible or where the action is so violent that all thermal measurements are impracticable.

Thus 28 grammes of olefiant gas, C_2H_4, contain 24 grammes of carbon and 4 grammes of hydrogen, and, although we can not combine directly charcoal and hydrogen gas, we can determine the heat evolved in the production of the compound in this way. If we burn 28 grammes of olefiant gas the products will be 88 grammes of carbonic dioxide and 36 grammes of water—

$$\overset{28}{C_2H_4} + 3O_2 = \overset{88}{2CO_2} + \overset{36}{2H_2O} = 332{,}024.$$

If we burn 24 grammes of charcoal and 4 grammes of hydrogen separately we shall obtain the same weights of the same products in the same condition—

$$\overset{24}{2C} + 2O_2 = \overset{88}{2CO_2} = 193{,}920$$
$$\overset{4}{2H_2} + O_2 = \overset{36}{2H_2O} = \underline{137{,}848}$$
$$331{,}768$$

The heat evolved in all these three processes of combustion has been measured and is given after the reaction. The total heat evolved in burning the elementary substances is less than that set free in burning the compound by 256 units. Hence in passing from charcoal and hydrogen gas to olefiant gas this small amount of heat must have been absorbed. Olefiant gas is therefore an endothermous compound.

Again, sulphuric oxide and water when united in the proportions indicated by the reaction—

$$SO_3 + H_2O = H_2SO_4$$

combine with explosive violence, but we can readily dissolve both SO_3 and H_2SO_4 in an equally large volume of water and determine the heat evolved in each case. The final result in both cases is a weak solution of sulphuric acid, and the difference between the amounts of heat evolved gives the amount which would be given by the above reaction if it could be measured.

(2) With a table giving the heats of formation of the more important compounds in different conditions (whether solid, liquid, aëriform, or in solution in water) we are in position to calculate the heat evolved in any ordinary chemical process. We have only to compare the sum of the heats of formation of the factors of the reaction with that of the products, paying careful regard to the conditions in which the several materials are present. Thus, in the reaction we have discussed so often—

$$(H_2SO_4 + Aq) + Zn = (ZnSO_4 + Aq) + H_2,$$

the heat evolved during the process is the difference between the heat of formation of sulphuric acid in aqueous solution and that of zinc sulphate dissolved in an equal amount of water. We need pay no regard to the elementary substances, either the zinc dissolved or the hydrogen gas set free, for they are present in the very condition which our calculations assume, and since the heat of formation of dilute sulphuric acid is 210,000 and that of the solution of zinc sulphate 252,000 there must be set free in the reaction—

$$252,000 - 210,000 = 42,000 \text{ units.}$$

(3) It has been inferred as a generalization from a great number of facts that, other things being equal, the activity of a chemical process is proportional to the amount of heat evolved, and that where several courses are possible the tendency is always to form those products which involve the greatest evolution of heat. Often by restraining the reaction (as by lowering the temperature, diluting the solution, or restricting the amount of material) other products may result; but if we give the chemical action full play the tendency is as above stated. In the action of nitric acid on copper there may be formed either nitric oxide or nitrogen gas, thus:

$$3Cu + (8HNO_3 + Aq) = (3Cu(NO_3)_2 + 4H_2O + Aq) + 2NO,$$

or

$$5Cu + (12HNO_3 + Aq) = (5Cu(NO_3)_2 + 6H_2O + Aq) + N_2.$$

The last develops the most heat, and nitrogen gas is the chief or sole product formed if the materials are allowed to become heated; but if the flask is kept cool the chief or only product is nitric oxide, as in Ex. 44 (*a*).

If no heat would be set free by an assumed process the reaction can not take place without some aid. There appears no reason in the form of the reaction why copper should not act on dilute sulphuric acid like zinc, that is—

$$Cu + (H_2SO_4 + Aq) \text{ yield } (CuSO_4 + Aq) + H_2$$

But while the heat of formation of an aqueous solution of sulphuric acid is 210,000, as above, that of an aqueous solution of copper sulphate is 199,100 (instead of 252,000, as in the case of zinc sulphate), and heat would be absorbed, not evolved, by the chemical change.

The aid required to determine a reaction in such cases may be furnished either by external energy, as the sun's rays acting on the green foliage of the vegetable kingdom or by some simultaneous exothermous process which entrains the other and supplies the necessary heat. Thus, in the above reaction, if a small amount of nitric acid is added (as shown in Ex. 54 (*b*)) the copper at once dissolves simply because the nitric acid, by oxidizing the hydrogen evolved (compare Ex. 43 (*b*)), generates the heat required to render the process, as a whole, exothermous.

Thus it is that the formation of endothermous compounds becomes possible. Nitrous oxide, N_2O, is such a compound. In its production from nitrogen and oxygen gases 18,000 units of heat are absorbed. Its tendency, therefore (in itself alone), is to fall back into the constituent gases, when the same amount of heat would be evolved. In the reaction by which nitrous oxide is made—

$$NH_4NO_2 = 2H_2O + N_2O$$
$$80,700 \quad\quad 118,800 - 18,000,$$

the heats of formation are printed under the symbol, and it will be seen that, as a whole, the process develops 100,800 units of heat. But, as will also be noticed, this heat wholly comes from the oxidation of hydrogen to form water, which is sufficient to furnish all that is required for the production of N_2O and still have a large excess over what is required to determine the reaction. Indeed, unless restrained (by keeping the temperature at the lowest possible point), this reaction will take the form—

$$NH_4NO_3 = 2H_2O + N_2 + \tfrac{1}{2}O_2,$$

which corresponds to a larger evolution of heat, and to just as much more as was used above in the production of N_2O.

(4) Endothermous compounds are always in a condition of unstable equilibrium, and sometimes highly explosive. This is strikingly true of iodide of nitrogen, which often explodes at the mere touch of a feather and is resolved wholly into elementary substances—

$$2NI_3 = N_2 + 3I_2.$$

They may endure, often for a long time, in consequence probably of features of molecular structure such as we endeavored to illustrate in the case of ammonium nitrate, but sooner or later they fall into more stable conditions. They may be compared to a vaulted cathedral roof of which the stones are firmly locked together and held high in air by buttresses, but when keystone or buttress fail fall in ruin.

There are many substances which, although exothermous to the elementary substances to which their heat of formation is referred, are endothermous in their relations to certain definite products into which they are more or less readily resolved with evolution of heat, or in their relations to associated material, from uniting with which they are restrained by physical disabilities or conditions of structure. Nitro-glycerin and gun-cotton, already referred to, are examples of the first type, while gunpowder, in which combustible charcoal and sulphur is kept apart from the great store of oxygen in the grains of nitre by the inertness of the solid state, is an equally striking example of the second type. All these explosive agents owe their efficiency not only to the heat evolved by the internal combustion, which ensues when they are fired, but also to the circumstance that the products into which they fall are for the most part aëriform bodies whose molecules acquire an enormous moving power under the influence of the heat thus generated.

Of instability arising from association by far the most wonderful example is furnished by the presence on the surface of the globe of a large amount of combustible material in contact with the oxygen of the atmosphere. Almost the whole of this material is made up either of the organized structure of plants and animals or else of the remains of such structures, and however multifarious the substances of which this organic matter may consist, the ultimate elements, with unimportant exceptions, are carbon, hydrogen, nitrogen, and oxygen, and the whole of this material was primarily formed from the constituents of air and water, including, of course, carbonic acid and ammonia, always present in the atmosphere and in the water permeating the soil. From these materials the plant obtains almost its sole food, and the animal ultimately at least lives on the plant. In some mysterious way the sun's rays give the plant the power of producing the substances of its tissues from the simple articles of its diet. Of the manner in which the highly complex products are built up we know almost nothing. But of this we are sure. The energy of the sun's rays is the power by which carbonic acid and water are decomposed and materials so unstable in the presence of the atmosphere constructed. Probably the effect is indeed a result of molecular construction and the structure endures until a conflagration, or the slower processes of decay, destroys the fabric and resolves the organic matter into the elements from which it sprung. Thus there is a constant cycle in nature. The sun's rays are ever building up, and in so doing are setting free the very oxygen which, sooner or later, will destroy all this work. We also know that the heat given out in burning is the exact equivalent of the work done in building, and therefore is simply transmuted solar energy. Just as the sun lifts the water whose fall maintains the great aqueous circulation of the globe, so the same vitalizing energy builds up organic structures, by whose reabsorption into the all-devouring atmosphere the life and activity on the planet is sustained. Moreover, only by utilizing this same energy—though often so indirectly that it escapes notice—are we able to produce endothermous or unstable compounds in our laboratories.

(5) It must be remembered that compounds are endothermous only in relation to the elementary substances in a definite condition (carbon as charcoal sulphur, as brimstone, etc.), from which in our system they are regarded as having been formed. Their peculiar thermal relations do not necessarily imply a want of chemical energy between the elementary atoms of which their molecules are supposed to consist. Thus the oxides of nitrogen are endothermous, and yet there can be no question that the nitrogen atoms have a marked affinity for the atoms of oxygen. If we could deal with elementary atoms all compounds would be, doubtless, exothermous, but when we deal with the elementary substances in their normal condition the constitution of the molecules of these elementary substances comes into play. As has been shown, the molecules of elementary substances, as well as those of compounds, are aggregates of atoms, only of atoms all of which are of the same kind, and not, as in compound substances, of different kinds. The molecule of oxygen gas (O_2) is formed of two atoms of oxygen ; that of ozone (O_3), of three atoms of oxygen ; that of nitrogen gas (N_2), of two atoms of nitrogen, etc. The atoms of nitrogen, united in a molecule of the gas must have an attraction for each other, else they would not so group themselves in pairs. As yet we have not been able to measure this attraction with confidence ; but we have good reason for believing that it is very strong. The endothermous compounds called nitrogen iodide (NI_3) and nitrogen chloride (NCl_3) are so explosive not, as we believe, because the nitrogen atoms have no affinity for the iodine and chlorine atoms, but because they have such a strong attraction for each other that they break from the iodine and chlorine atoms and rush together.

$$2NI_3 = N_2 + 3I_2$$
$$2NCl_3 = N_2 + 3Cl_2$$

We explain the inertness of nitrogen gas in this way. Simply the nitrogen atoms exert a stronger attraction among themselves than for those of the other chemical elements with only a few exceptions. Hence, also, the general instability of the compounds of nitrogen as a class. Animal structures are for the most part made up of nitrogenized substance, and thus decay and death in nature are closely associated with this striking feature of our chemical philosophy.

(6) Could we experiment with isolated atoms all chemical relations would unquestionably appear simpler, and modern science has rendered probable that there exists such a condition in the universe. It is well known that heat tends to decompose chemical compounds, and the phenomena thus resulting form a very interesting subject of chemical inquiry known as thermolysis or dissociation. Steam passed through metal tubes at a white heat acts in every respect like a mixture of oxygen and hydrogen gases, and, as common experience shows, most compounds even at a red heat suffer more or less fundamental chemical changes. The known facts point to the conclusion that at such high temperatures as must rule at the sun and at the fixed stars all known materials would be resolved into elementary atoms. Spectroscopic observations confirm this inference, and such evidence has even been interpreted as indicating that some of the atoms which we regard as elementary are resolved into still simpler parts at the great focus of solar radiation. If the nebular hypothesis is correct our world must have been primarily in this condition, and the substances which we now find on its crust must have been formed as the elementary atoms came together in the process of cooling and united in accordance with their mutual affinities. According to this hypothesis, the original chaos out of which the present order sprang was a condition of isolated atoms, and the foundations of the globe must have been laid in flames.

(7) In this last chapter of our book we have only been able to illustrate a few of the more important relations of thermo-chemistry. Our one object has been to exhibit the scope of this department of chemical science, as we have sought to show in earlier chapters that of qualitative and

of quantitative analysis. The theoretical relations of the science have been previously set forth by us in popular form in the New Chemistry.* It is expected that this will serve as a companion to the former book, and the student who thoughtfully performs all the experiments and diligently inquires what each is calculated to teach can not fail to gain clear ideas of the methods of chemical investigation and at the same time will acquire skill in drawing inferences from experimental data. Having thus seen the relations of the broader divisions of the subject, the student will be prepared to enter on the professional study of chemistry intelligently, or, if he goes no further, will have acquired a clear conception of the aims and methods of the science and of its true position in a scheme of education. The next step in the study of chemistry should be to acquire an adequate knowledge of the scheme of the chemical elements as it is presented by Roscoe and Schorlemmer,† or, still better, as it is illustrated experimentally in an extended course of lectures such as is given every year at Harvard or at any one of our principal universities. This is a serious task, since the mass of details is very great and the subject can not be profitably abridged beyond a limited extent. A brief epitome will not be of much value. The mind must dwell on the subject in its various relations in order to make the knowledge real or lasting, and unless the student has that object in the acquisition which will lead him to give to the work the requisite time, he had better, especially if it is a question of liberal education, limit his study to the general principles of the science as presented in this or some similar book.

---

* D. Appleton & Co., publishers. New York. Last edition, 1890.

† Treatise on Chemistry, D. Appleton & Co.

List of the Elementary Substances, excepting such as are very Rare or of Doubtful Authenticity.

Aluminum,	Al,	27·1	Molybdenum,	Mo,		96
Antimony,	Sb,	120	Nickel,	Ni,		59
Arsenic,	As,	75	Nitrogen,	N,		14
Barium,	Ba,	137	Osmium,	Os,		190·9
Bismuth,	Bi,	208	Oxygen,	O,		16
Boron,	B,	11	Palladium,	Pd,		106·6
Bromine,	Br,	80	Phosphorus,	P,		31
Cadmium,	Cd,	112·2	Platinum,	Pt,		194·8
Cæsium,	Cs,	133	Potassium,	K,		39·1
Calcium,	Ca,	40	Rhodium,	Rh,		103
Carbon,	C,	12	Rubidium,	Rb,		85·4
Cerium,	Ce,	140	Ruthenium,	Ru,		101·8
Chlorine,	Cl,	35·5	Scandium,	Sc,		44
Chromium,	Cr,	52·1	Selenium,	Se,		79·2
Cobalt,	Co,	59	Silicon,	Si,		28·3
Columbium,	Cb,	94	Silver,	Ag,		108
Copper,	Cu,	63·6	Sodium,	Na,		23
Didymium,	Nd, 141	Pr, 144	Strontium,	Sr,		87·6
Erbium,	Er,	166	Sulphur,	S,		32·1
Fluorine,	F,	19	Tantalum,	Ta,		182
Gallium,	Ga,	70	Tellurium,	Te,		125 ?
Germanium,	Ge,	72·3	Terbium ?	Tr,		171 ?
Glucinum,	Gl,	9·1	Thallium,	Tl,		204·1
Gold,	Au,	197·2	Thorium,	Th,		232
Hydrogen,	H,	1	Thulium ?	Tm,		171
Indium,	In,	113·7	Tin,	Sn,		118
Iodine,	I,	126·9	Titanium,	Ti,		48
Iridium,	Ir,	193	Tungsten,	W,		184
Iron,	Fe,	56	Uranium,	Ur,		240
Lanthanum,	La,	139	Vanadium,	Va,		51·4
Lead,	Pb,	206·9	Ytterbium ?	Yb,		173
Lithium,	Li,	7	Yttrium,	Y,		90
Magnesium,	Mg,	24·4	Zinc,	Zn,		65·2
Manganese,	Mn,	55	Zirconium,	Zr,		90
Mercury,	Hg,	200				

THE END.

KEROSENE STOVE AND TUBE FURNACE.
Well adapted for most chemical experiments referred to in this book (Ex. 23).

D. APPLETON & CO.'S PUBLICATIONS.

JOHN TYNDALL'S WORKS.

ESSAYS ON THE FLOATING MATTER OF THE AIR, in Relation to Putrefaction and Infection. 12mo. Cloth, $1.50.

ON FORMS OF WATER, in Clouds, Rivers, Ice, and Glaciers. With 35 Illustrations. 12mo. Cloth, $1.50.

HEAT AS A MODE OF MOTION. New edition. 12mo. Cloth, $2.50.

ON SOUND: A Course of Eight Lectures delivered at the Royal Institution of Great Britain. Illustrated. 12mo. New edition Cloth, $2.00.

FRAGMENTS OF SCIENCE FOR UNSCIENTIFIC PEOPLE. 12mo. New revised and enlarged edition. Cloth, $2.50.

LIGHT AND ELECTRICITY. 12mo. Cloth, $1.25.

LESSONS IN ELECTRICITY, 1875–'76. 12mo. Cloth, $1.00.

HOURS OF EXERCISE IN THE ALPS. With Illustrations. 12mo. Cloth, $2.00.

FARADAY AS A DISCOVERER. A Memoir. 12mo. Cloth, $1.00.

CONTRIBUTIONS TO MOLECULAR PHYSICS in the Domain of Radiant Heat. $5.00.

SIX LECTURES ON LIGHT. Delivered in America in 1872–'73. With an Appendix and numerous Illustrations. Cloth, $1.50

ADDRESS delivered before the British Association, assembled at Belfast. Revised with Additions. 12mo. Paper, 50 cents.

RESEARCHES ON DIAMAGNETISM AND MAGNE-CRYSTALLIC ACTION, including the Question of Diamagnetic Polarity. With Ten Plates. 12mo, cloth. Price, $1.50.

New York: D. APPLETON & CO., 1, 3, & 5 Bond Street

D. APPLETON & CO.'S PUBLICATIONS.

GEORGE J. ROMANES'S WORKS.

MENTAL EVOLUTION IN MAN: Origin of Human Faculty. One vol., 8vo. Cloth, $3.00.

This work, which follows "Mental Evolution in Animals," by the same author, considers the probable mode of genesis of the human mind from the mind of lower animals, and attempts to show that there is no distinction of kind between man and brute, but, on the contrary, that such distinctions as do exist all admit of being explained, with respect to their evolution, by adequate psychological analysis.

"The vast array of facts, and the sober and solid method of argument employed by Mr. Romanes, will prove, we think, a great gift to knowledge."—*Saturday Review.*

JELLY-FISH, STAR-FISH, AND SEA-URCHINS. Being a Research on Primitive Nervous Systems. 12mo. Cloth, $1.75.

"Although I have throughout kept in view the requirements of a general reader, I have also sought to render the book of service to the working physiologist, by bringing together in one consecutive account all the more important observations and results which have been yielded by this research."—*Extract from Preface.*

"A profound research into the laws of primitive nervous systems conducted by one of the ablest English investigators. Mr. Romanes set up a tent on the beach and examined his beautiful pets for six summers in succession. Such patient and loving work has borne its fruits in a monograph which leaves nothing to be said about jelly-fish, star-fish, and sea-urchins. Every one who has studied the lowest forms of life on the sea-shore admires these objects. But few have any idea of the exquisite delicacy of their structure and their nice adaptation to their place in nature. Mr. Romanes brings out the subtile beauties of the rudimentary organisms, and shows the resemblances they bear to the higher types of creation. His explanations are made more clear by a large number of illustrations."—*New York Journal of Commerce.*

ANIMAL INTELLIGENCE. 12mo. Cloth, $1.75.

"A collection of facts which, though it may merely amuse the unscientific reader, will be a real boon to the student of comparative psychology, for this is the first attempt to present systematically the well-assured results of observation on the mental life of animals."—*Saturday Review.*

MENTAL EVOLUTION IN ANIMALS. With a Posthumous Essay on Instinct, by CHARLES DARWIN. 12mo. Cloth, $2.00.

"Mr. Romanes has followed up his careful enumeration of the facts of 'Animal Intelligence,' contributed to the 'International Scientific Series,' with a work dealing with the successive stages at which the various mental phenomena appear in the scale of life. The present installment displays the same evidence of industry in collecting facts and caution in co-ordinating them by theory as the former"—*The Athenæum.*

New York: D. APPLETON & CO., 1, 3, & 5 Bond Street.

D. APPLETON & CO.'S PUBLICATIONS.

ERNST HAECKEL'S WORKS.

THE HISTORY OF CREATION; OR, THE DEVELOPMENT OF THE EARTH AND ITS INHABITANTS BY THE ACTION OF NATURAL CAUSES. A Popular Exposition of the Doctrine of Evolution in general, and of that of Darwin, Goethe, and Lamarck in particular. From the German of ERNST HAECKEL, Professor in the University of Jena. The translation revised by Professor E. Ray Lankester, M. A., F. R. S., Fellow of Exeter College, Oxford. Illustrated with Lithographic Plates. In two vols., 12mo. Cloth, $5.00.

THE EVOLUTION OF MAN. A Popular Exposition of the Principal Points of Human Ontogeny and Phylogeny. From the German of ERNST HAECKEL, Professor in the University of Jena, author of "The History of Creation," etc. With numerous Illustrations. In two vols., 12mo. Cloth. Price, $5.00.

"In this excellent translation of Professor Haeckel's work, the English reader has access to the latest doctrines of the Continental school of evolution, in its application to the history of man. It is in Germany, beyond any other European country, that the impulse given by Darwin twenty years ago to the theory of evolution has influenced the whole tenor of philosophical opinion. There may be, and are, differences in the degree to which the doctrine may be held capable of extension into the domain of mind and morals; but there is no denying, in scientific circles at least, that as regards the physical history of organic nature much has been done toward making good a continuous scheme of being."
—*London Saturday Review.*

FREEDOM IN SCIENCE AND TEACHING. From the German of ERNST HAECKEL. With a Prefatory Note by T. H. HUXLEY, F. R. S. 12mo. $1.00.

New York: D. APPLETON & CO., 1, 3, & 5 Bond Street.

Professor JOSEPH LE CONTE'S WORKS.

EVOLUTION AND ITS RELATION TO RELIGIOUS THOUGHT. By JOSEPH LE CONTE, LL. D., Professor of Geology and Natural History in the University of California. With numerous Illustrations. 12mo. Cloth, $1.50.

"Much, very much has been written, especially on the nature and the evidences of evolution, but the literature is so voluminous, much of it so fragmentary, and most of it so technical, that even very intelligent persons have still very vague ideas on the subject. I have attempted to give (1) a very concise account of what we mean by evolution, (2) an outline of the evidences of its truth drawn from many different sources, and (3) its relation to fundamental religious beliefs."
—*Extract from Preface.*

ELEMENTS OF GEOLOGY. A Text-book for Colleges and for the General Reader. By JOSEPH LE CONTE, LL. D. With upward of 900 Illustrations. New and enlarged edition. 8vo. Cloth, $4.00.

"Besides preparing a comprehensive text-book, suited to present demands, Professor Le Conte has given us a volume of great value as an exposition of the subject, thoroughly up to date. The examples and applications of the work are almost entirely derived from this country, so that it may be properly considered an American geology. We can commend this work without qualification to all who desire an intelligent acquaintance with geological science, as fresh, lucid, full, authentic, the result of devoted study and of long experience in teaching."
—*Popular Science Monthly.*

RELIGION AND SCIENCE. A Series of Sunday Lectures on the Relation of Natural and Revealed Religion, or the Truths revealed in Nature and Scripture. By JOSEPH LE CONTE, LL. D. 12mo. Cloth, $1.50.

"We commend the book cordially to the regard of all who are interested in whatever pertains to the discussion of these grave questions, and especially to those who desire to examine closely the strong foundations on which the Christian faith is reared."—*Boston Journal.*

SIGHT: An Exposition of the Principles of Monocular and Binocular Vision. By JOSEPH LE CONTE, LL. D. With Illustrations. 12mo. Cloth, $1.50.

"Professor Le Conte has long been known as an original investigator in this department; all that he gives us is treated with a master-hand. It is pleasant to find an American book that can rank with the very best of foreign books on this subject."—*The Nation.*

New York: D. APPLETON & CO., 1, 3, & 5 Bond Street.

D. APPLETON & CO.'S PUBLICATIONS.

ALEXANDER BAIN'S WORKS.

THE SENSES AND THE INTELLECT. 8vo. Cloth, $5.00.

The object of this treatise is to give a full and systematic account of two principal divisions of the science of mind—the senses and the intellect. The value of the third edition of the work is greatly enhanced by an account of the psychology of Aristotle, which has been contributed by Mr. Grote.

THE EMOTIONS AND THE WILL. 8vo. Cloth, $5.00.

The present publication is a sequel to the former one on "The Senses and the Intellect," and completes a systematic exposition of the human mind.

MIND AND BODY. Theories of their Relations. 12mo. Cloth, $1.50.

"A forcible statement of the connection between mind and body, studying their subtile interworkings by the light of the most recent physiological investigations."—*Christian Register.*

EDUCATION AS A SCIENCE. 12mo. Cloth, $1.75.

ON TEACHING ENGLISH. With Detailed Examples and an Inquiry into the Definition of Poetry. 12mo. Cloth, $1.25.

PRACTICAL ESSAYS. 12mo. Cloth, $1.50.

Dr. H. ALLEYNE NICHOLSON'S WORKS.

MANUAL OF ZOÖLOGY, for the Use of Students, with a General Introduction to the Principles of Zoölogy. Second edition. Revised and enlarged, with 243 Woodcuts. 12mo. Cloth, $2.50.

THE ANCIENT LIFE-HISTORY OF THE EARTH. A Comprehensive Outline of the Principles and Leading Facts of Palæontological Science. 12mo. Cloth, $2.00.

"A work by a master in the science who understands the significance of every phenomenon which he records, and knows how to make it reveal its lessons. As regards its value there can scarcely exist two opinions. As a text-book of the historical phase of palæontology it will be indispensable to students, whether specially pursuing geology or biology; and without it no man who aspires even to an outline knowledge of natural science can deem his library complete."—*The Quarterly Journal of Science.*

New York: D. APPLETON & CO., 1, 3, & 5 Bond Street.

D. APPLETON & CO.'S PUBLICATIONS.

THOMAS H. HUXLEY'S WORKS.

SCIENCE AND CULTURE, AND OTHER ESSAYS. 12mo. Cloth, $1.50.

THE CRAYFISH: AN INTRODUCTION TO THE STUDY OF ZOÖLOGY. With 82 Illustrations. 12mo. Cloth, $1.75.

MAN'S PLACE IN NATURE. 12mo. Cloth, $1.25.

ON THE ORIGIN OF SPECIES. 12mo. Cloth, $1.00.

MORE CRITICISMS ON DARWIN, AND ADMINISTRATIVE NIHILISM. 12mo. Limp cloth, 50 cents.

MANUAL OF THE ANATOMY OF VERTEBRATED ANIMALS. Illustrated. 12mo. Cloth, $2.50.

MANUAL OF THE ANATOMY OF INVERTEBRATED ANIMALS. 12mo. Cloth, $2.50.

LAY SERMONS, ADDRESSES, AND REVIEWS. 12mo. Cloth, $1.75.

CRITIQUES AND ADDRESSES. 12mo. Cloth, $1.50.

AMERICAN ADDRESSES; WITH A LECTURE ON THE STUDY OF BIOLOGY. 12mo. Cloth, $1.25.

PHYSIOGRAPHY: AN INTRODUCTION TO THE STUDY OF NATURE. With Illustrations and Colored Plates. 12mo. Cloth, $2.50.

THE ADVANCE OF SCIENCE IN THE LAST HALF-CENTURY. 12mo. Paper, 25 cents.

New York: D. APPLETON & CO., 1, 3, & 5 Bond Street.

D. APPLETON & CO.'S PUBLICATIONS.

SIR JOHN LUBBOCK'S (Bart.) WORKS.

THE ORIGIN OF CIVILIZATION AND THE PRIMITIVE CONDITION OF MAN, MENTAL AND SOCIAL CONDITION OF SAVAGES. Fourth edition, with numerous Additions. With Illustrations. 8vo. Cloth, $5.00.

"This interesting work—for it is intensely so in its aim, scope, and the ability of its author—treats of what the scientists denominate *anthropology*, or the natural history of the human species; the complete science of man, body, and soul, including sex, temperament, race, civilization, etc."—*Providence Press.*

PREHISTORIC TIMES, AS ILLUSTRATED BY ANCIENT REMAINS AND THE MANNERS AND CUSTOMS OF MODERN SAVAGES. Illustrated. 8vo. Cloth, $5.00.

"This is, perhaps, the best summary of evidence now in our possession concerning the general character of prehistoric times. The Bronze Age, The Stone Age, The Tumuli, The Lake Inhabitants of Switzerland, The Shell Mounds, The Cave Man, and The Antiquity of Man, are the titles of the most important chapters."—*Dr. C. K. Adams's Manual of Historical Literature.*

ANTS, BEES, AND WASPS. A Record of Observations on the Habits of the Social Hymenoptera. With Colored Plates. 12mo. Cloth, $2.00.

"This volume contains the record of various experiments made with ants, bees, and wasps during the last ten years, with a view to test their mental condition and powers of sense. The author has carefully watched and marked particular insects, and has had their nests under observation for long periods—one of his ants' nests having been under constant inspection ever since 1874. His observations are made principally upon ants, because they show more power and flexibility of mind; and the value of his studies is that they belong to the department of original research."

ON THE SENSES, INSTINCTS, AND INTELLIGENCE OF ANIMALS, WITH SPECIAL REFERENCE TO INSECTS. "International Scientific Series." With over One Hundred Illustrations. 12mo. Cloth, $1.75.

The author has here collected some of his recent observations on the senses and intelligence of animals, and especially of insects, and has attempted to give, very briefly, some idea of the organs of sense, commencing in each case with those of man himself.

THE PLEASURES OF LIFE. 12mo. Cloth, 50 cents; paper, 25 cents.

CONTENTS.—THE DUTY OF HAPPINESS. THE HAPPINESS OF DUTY. A SONG OF BOOKS. THE CHOICE OF BOOKS. THE BLESSING OF FRIENDS. THE VALUE OF TIME. THE PLEASURES OF TRAVEL. THE PLEASURES OF HOME. SCIENCE. EDUCATION.

New York: D. APPLETON & CO., 1, 3, & 5 Bond Street.

D. APPLETON & CO.'S PUBLICATIONS.

DR. HENRY MAUDSLEY'S WORKS.

BODY AND WILL: Being an Essay concerning Will in its Metaphysical, Physiological, and Pathological Aspects. 12mo. Cloth, $2.50.

BODY AND MIND: An Inquiry into their Connection and Mutual Influence, specially in reference to Mental Disorders. 1 vol., 12mo. Cloth, $1.50.

PHYSIOLOGY AND PATHOLOGY OF MIND:

PHYSIOLOGY OF THE MIND. New edition. 1 vol., 12mo Cloth, $2.00. CONTENTS: Chapter I. On the Method of the Study of the Mind.—II. The Mind and the Nervous System.—III. The Spinal Cord, or Tertiary Nervous Centres; or, Nervous Centres of Reflex Action.—IV. Secondary Nervous Centres; or, Sensory Ganglia; Sensorium Commune.—V. Hemispherical Ganglia; Cortical Cells of the Cerebral Hemispheres; Ideational Nervous Centres, Primary Nervous Centres; Intellectorium Commune.—VI. The Emotions.—VII. Volition.—VIII.—Motor Nervous Centres, or Motorium Commune and Actuation or Effection.—IX. Memory and Imagination.

PATHOLOGY OF THE MIND. Being the Third Edition of the Second Part of the "Physiology and Pathology of Mind," recast, enlarged, and rewritten. 1 vol., 12mo. Cloth, $2.00. CONTENTS: Chapter I. Sleep and Dreaming.—II. Hypnotism, Somnambulism, and Allied States.—III. The Causation and Prevention of Insanity: (A) Etiological.—IV. The same continued.—V. The Causation and Prevention of Insanity: (B) Pathological.—VI. The Insanity of Early Life.—VII. The Symptomatology of Insanity.—VIII. The same continued.—IX. Clinical Groups of Mental Disease.—X. The Morbid Anatomy of Mental Derangement.—XI. The Treatment of Mental Disorders.

RESPONSIBILITY IN MENTAL DISEASE. (International Scientific Series.) 1 vol., 12mo. Cloth, $1.50.

"The author is at home in his subject, and presents his views in an almost singularly clear and satisfactory manner. . . . The volume is a valuable contribution to one of the most difficult and at the same time one of the most important subjects of investigation at the present day."—*New York Observer.*

"Handles the important topic with masterly power, and its suggestions are practical and of great value."—*Providence Press.*

New York: D. APPLETON & CO., 1, 3, & 5 Bond Street.

D. APPLETON & CO.'S PUBLICATIONS.

Professor E. L. YOUMANS'S WORKS.

THE HAND-BOOK OF HOUSEHOLD SCIENCE. A Popular Account of Heat, Light, Air, Aliment, and Cleansing, in their Scientific Principles and Domestic Applications. 12mo. Illustrated. Cloth, $1.75.

THE CULTURE DEMANDED BY MODERN LIFE. A Series of Addresses and Arguments on the Claims of Scientific Education. Edited, with an Introduction on Mental Discipline in Education. 1 vol., 12mo. Cloth, $2.00.

CORRELATION AND CONSERVATION OF FORCES. A Series of Expositions by Scientific Men. Edited, with an Introduction and Brief Biographical Notices of the Chief Promoters of the New Views, by EDWARD L. YOUMANS, M. D. 12mo. Cloth, $2.00.

CONTENTS.

I. By Professor W. R. GROVE. The Correlation of Physical Forces.
II. By Professor HELMHOLTZ. The Interaction of Natural Forces.
III. By Dr. J. R. MAYER. 1. Remarks on the Forces of Inorganic Nature.
    2. On Celestial Dynamics.
    3. On the Mechanical Equivalent of Heat.
IV. By Dr. FARADAY. Some Thoughts on the Conservation of Forces.
V. By Professor LIEBIG. The Connection and Equivalence of Forces.
VI. By Dr. CARPENTER. The Correlation of the Physical and Vital Forces.

"This work is a very welcome addition to our scientific literature, and will be particularly acceptable to those who wish to obtain a popular but at the same time precise and clear view of what Faraday justly calls the highest law in physical science, the principle of the conservation of force. Sufficient attention has not been paid to the publication of collected monographs or memoirs upon special subjects. Dr. Youmans's work exhibits the value of such collections in a very striking manner, and we earnestly hope his excellent example may be followed in other branches of science." *American Journal of Science.*

New York: D. APPLETON & CO., 1, 3, & 5 Bond Street.

THE POPULAR SCIENCE MONTHLY.

Established by Edward L. Youmans.

EDITED BY W. J. YOUMANS.

The Popular Science Monthly will continue, as heretofore, to supply its readers with the results of the latest investigation and the most valuable thought in the various departments of scientific inquiry.

Leaving the dry and technical details of science, which are of chief concern to specialists, to the journals devoted to them, the Monthly deals with those more general and practical subjects which are of the greatest interest and importance to the public at large. In this work it has achieved a foremost position, and is now the acknowledged organ of progressive scientific ideas in this country.

The wide range of its discussions includes, among other topics:

The bearing of science upon education;

Questions relating to the prevention of disease and the improvement of sanitary conditions;

Subjects of domestic and social economy, including the introduction of better ways of living, and improved applications in the arts of every kind;

The phenomena and laws of the larger social organizations, with the new standard of ethics, based on scientific principles;

The subjects of personal and household hygiene, medicine, and architecture, as exemplified in the adaptation of public buildings and private houses to the wants of those who use them;

Agriculture and the improvement of food-products;

The study of man, with what appears from time to time in the departments of anthropology and archæology that may throw light upon the development of the race from its primitive conditions.

Whatever of real advance is made in chemistry, geography, astronomy, physiology, psychology, botany, zoölogy, paleontology, geology, or such other department as may have been the field of research, is recorded monthly.

Special attention is also called to the biographies, with portraits, of representative scientific men, in which are recorded their most marked achievements in science, and the general bearing of their work indicated and its value estimated.

Terms: $5.00 per annum, in advance.

The New York Medical Journal and The Popular Science Monthly to the same address, $9.00 per annum (full price, $10.00).

New York: D APPLETON & CO., 1, 3, & 5 Bond Street.

www.ingramcontent.com/pod-product-compliance
Lightning Source LLC
Chambersburg PA
CBHW020909230426
43666CB00008B/1377